NATURE'S CLEAN-UP CREW: THE IMPACT OF MICROBIAL TECHNOLOGY ON WASTE REMEDIATION

طاقم تنظيف الطبيعة: تأثير التكنولوجيا الميكروبية على معالجة النفايات

اللهم منك وإليك

FROM AND TO ALLAH

Authors

Dr. Tarek H. Taha is a Professor at the Environmental Biotechnology Department, Genetic Engineering and Biotechnology Research Institute (GEBRI), City of Scientific Research and Technological Applications (SRTA-CITY), Alexandria, Egypt. He was a visiting Professor at Newcastle University, UK. He has his expertise in the field of Environmental Biotechnology. His research interest is concerned by the Biomonitoring and Bioremediation of environmental contaminants. He is also interested in the biosynthesis of nanoparticles and their applications in biosensors and other environmental fields, and has a great passion with Bioinformatics, Molecular techniques, and Genetic engineering. In addition, he is interested in the production of biofuel from environmental wastes, and finally, the bioconversion of environmental wastes into value-added products.

Dr. Ahmed K. Saleh is a Researcher at Cellulose and Paper Department, Chemical Industrious Research Institute, National Research Centre (Egypt). His current research focuses on the sustainable production of bacterial cellulose and bioethanol. Other expertise includes application of bacterial cellulose, green synthesis of nanoparticles, experimental design, and valorization of environmental and agricultural wastes.

Dr. Gamal A. G. Ammar is an Associate Professor of Biotechnology at SRTA-City, Egypt. His professional career started by working for prestigious organizations like FAO and Monsanto, and as a visiting scientist for top Universities like; Oxford, Cambridge, UK; Paris University, France; and Cornell University, USA. Ammar is the research team leader of a distinguished group experienced in the biotechnology industry with different activities and patents for utilizing the wide biotechnology umbrella, by employing plant systems in environmental aspects like; bioremediation, and plant-microbe interaction; medical ones like producing edible vaccines and secondary metabolites. He is also skilled in Genome Editing, Plant Tissue Culture and Transformation, Environmental Biotechnology, Molecular Biology, and Microbial genetics.

Dr. Abdulrahman M. Alhudhaibi is an Assistant Professor in the department of Biology, College of Science, Imam Mohammed ibn Saud Islamic University. His research interests lie in the field of development of antimicrobial products.

Dr. Feras Mohammed Afifi, holding a PhD degree from Newcastle University, UK. He is working at Umm Al- Qura University at Saudi Arabia. His research interested on Agricultural Biotechnology like, development of novel biopesticide systems using dsRNA molecules to control major crop pest via the natural RNA interference (RNAi) response.

Dr. Abdullah A. Faqihi is an Assistant Professor in the Department of Industrial Engineering, College of Engineering and Computer Science, Jazan University, Jazan, Kingdom of Saudi Arabia. His interests lie in the fields of manufacturing processes, materials science, and the use of graphene oxide to produce medical biosensors.

Dr. Asmaa Aboushaishaa has graduated from the faculty of medicine and surgery, 6th October University in 2014. She has recently completed her MRCPCH in 2023, in addition to obtaining IBCLC in 2021, and now she is working as a pediatrician and a lactation conslatant.

Contents

English Title	Page Number رقم الصفحة	العنوان العربي
Abstract	14	الملخص العربي
Keywords	15	الكلمات المفتاحية
1. Introduction	16	1. المقدمة
2. Definition and importance of bioremediation	18	2. تعريف المعالجة الحيوية وأهميتها
2.1. In-situ bioremediation	23	2.1. المعالجة الحيوية داخل الموقع
2.2. Ex-situ bioremediation	25	2.2. المعالجة الحيوية خارج الموقع
3. Microbial Technologies in Bioremediation	29	3. التقنيات الميكروبية في المعالجة الحيوية
3.1. Enzymatic oxidation	31	3.1. الأكسدة الإنزيمية
3.2. Enzymatic reduction	31	3.2. الإختزال الإنزيمى
3.3. Bioaugmentation	32	3.3. التعزيز الحيوي
3.4. Biostimulation	34	3.4. التحفيز الحيوي
3.5. Bioleaching	35	3.5. التصفية الحيوية
3.6. Biosorption	36	3.6. الإمتصاص الحيوي
3.7. Bioaccumulation	38	3.7. التراكم الحيوي
3.8. Precipitation	39	3.8. الترسيب
4. Using of microbial enzymes for multipollutant bioremediation	41	4. إستخدام الإنزيمات الميكروبية للمعالجة الحيوية للملوثات المتعددة
4.1. Hydrolytic enzymes	42	4.1 الإنزيمات المحللة

4.2. Lipases	44	4.2. الليبيزيز
4.3. Oxidoreductases	45	4.3. إنزيمات الأكسدة والاختزال
4.4. Oxygenases	46	4.4. إنزيمات الأكسجين
4.4.1. Monooxygenases	47	4.4.1. إنزيمات الأكسجين الأحادية
4.4.2. Dioxygenases	48	4.4.2. إنزيمات الأكسجين الثنائية
4.5. Laccases	49	4.5. اللاكيزيز
4.6. Cytochrome P450	51	4.6. سيتوكروم P450
5. Genetic modification and engineering for enhanced bioremediation	53	5. التعديل الوراثي والهندسة الوراثية لتعزيز المعالجة الحيوية
6. Types of Environmental Wastes	58	6. أنواع النفايات البيئية
6.1. Wastewater	58	6.1. مياه الصرف
6.1.1. Factors cause water pollution	62	6.1.1. العوامل المسببة لتلوث المياه
6.1.1.1. Inorganic compounds	62	6.1.1.1. المركبات غير العضوية
6.1.1.2. Organic compounds	63	6.1.1.2. المركبات العضوية
6.1.2. Microbial-assisted nanomaterials for wastewater treatment	64	6.1.2. المواد النانوية بمساعدة الميكروبات لمعالجة مياه الصرف
6.1.3. Enzyme-integrated nanoparticles for wastewater treatment	67	6.1.3. جسيمات نانوية مدمجة بالإنزيمات لمعالجة مياه الصرف

6.2. Solid waste	69	6.2. المخلفات الصلبة
6.3. Food waste and agricultural residues	71	6.3. المخلفات الغذائية والبقايا الزراعية
6.3.1. Bioremediation of domestic food waste	73	6.3.1. المعالجة الحيوية لمخلفات الطعام المنزلية
6.3.2. Bioremediation of wastes from food processing industries	76	6.3.2. المعالجة الحيوية للمخلفات الناتجة عن الصناعات الغذائية
6.3.3. Bioremediation of catering wastes	77	6.3.3. المعالجة الحيوية لمخلفات المطاعم
6.4. Industrial waste	79	6.4. المخلفات الصناعية
6.4.1. Organic industrial wastes	79	6.4.1. المخلفات الصناعية العضوية
6.4.2. Inorganic industrial wastes	79	6.4.2. المخلفات الصناعية الغير عضوية
6.4.3. Hazardous industrial waste	80	6.4.3. المخلفات الصناعية الخطرة
6.4.4. Non-hazardous industrial waste	80	6.4.4. المخلفات الصناعية الغير خطرة
6.4.5. Mining industrial waste	80	6.4.5. مخلفات صناعة التعدين
6.4.6. Metallurgical industrial waste	81	6.4.6. مخلفات صناعة المعادن
6.4.7. Chemical industrial waste	81	6.4.7. المخلفات الصناعية الكيميائية
6.4.8. Food preservation industrial waste	81	6.4.8. المخلفات الصناعية من حفظ الاغذية

6.4.9. Constructional materials industrial waste	81	6.4.9. المخلفات الصناعية من مواد البناء
6.4.10. Fired industrial waste	82	6.4.10. حرق المخلفات الصناعية
6.4.11. Unfired industrial waste	82	6.4.11. المخلفات الصناعية غير المشتعلة
6.5. Medical waste	82	6.5. المخلفات الطبية
6.5.1. The environmental impact of medical wastes	85	6.5.1. التأثير البيئي للمخلفات الطبية
6.5.2. Biotechnological managing of medical wastes	86	6.5.2. الإدارة التكنولوجية الحيوية للمخلفات الطبية
6.5.2.1. Bioremediation of medical wastes	87	6.5.2.1. المعالجة الحيوية للمخلفات الطبية
6.5.2.2. Composting of medical wastes	89	6.5.2.2. تسميد المخلفات الطبية
6.5.2.3. Phytoremediation of medical wastes	91	6.5.2.3. المعالجة النباتية للمخلفات الطبية
6.6. Electronic wastes (e-waste)	93	6.6. المخلفات الإلكترونية (النفايات الإلكترونية)
6.6.1. Biotechnological recycling of e-waste	95	6.6.1. إعادة تدوير المخلفات الإلكترونية باستخدام التكنولوجيا الحيوية

6.6.2. Potential biodegradation of e-waste plastics	99	6.6.2. التحلل الحيوي الفعال للمخلفات البلاستيكية الإلكترونية
7. Microbial metabolites used for bioremediation	102	7. المنتجات الايضية الميكروبية المستخدمة في المعالجة الحيوية
8. Production of value-added products from waste streams	103	8. إنتاج منتجات ذات قيمة من مجاري المخلفات
9. Energy production via microbial fuel cell (MFC)	107	9. إنتاج الطاقة عن طريق خلايا الوقود الميكروبية (MFC)
9.1. Principles of MFCs	107	9.1. مبادئ MFCs
9.2. Environmental wastes as substrates for MFC systems	108	9.2. المخلفات البيئية كركائز لأنظمة MFC
10. Latest advances in microbial bioremediation	114	10. أحدث التطورات في المعالجة الحيوية الميكروبية
10.1. Microbial glycoconjugates	114	10.1. الجليكوكونجات الميكروبية
10.2. Microbial biofilm	115	10.2. الأغشية الحيوية الميكروبية
10.3. Bioelectrochemical system	115	10.3. النظام الكهروكيميائي الحيوي
10.4. Nanoparticles	116	10.4. الجسيمات النانوية
11. Challenges, future perspectives, and recommendations	117	11. التحديات والآفاق المستقبلية والتوصيات
12. Egyptian experiences and	121	12. خبرة ومشاركة المصرين في مجال

participations in the field of waste's bioremediation		المعالجة الحيوية للمخلفات
13. The Role of Kingdom of Saudi Arabia in Bioremediation Technology	133	13. دور المملكة العربية السعودية في تكنولوجيا المعالجة البيولوجية
14. The role of some other Arab countries in bioremediation technology	135	14. دور بعض الدول العربية الأخرى في تكنولوجيا المعالجة البيولوجية
15. The scientific and applied benefits	138	15. الفوائد العلمية والتطبيقية
16. Conclusion	141	16. الخاتمة
17. References	142	17. المراجع

Nature's Clean-Up Crew: The Impact of Microbial Technology on Waste Remediation
طاقم تنظيف الطبيعة: تأثير التكنولوجيا الميكروبية على معالجة النفايات

Abstract

The daily disposal of wastes into the environment is a threatening issue that all countries are consistently facing. Due to the harmful influences of such wastes on humans, animals, and the ecosystem; a lot of techniques have been developed to deal with the toxicity resulting from such wastes. Although the chemical and physical treatments showed effectiveness regarding the remediation of these pollutants, they are expensive techniques that lacking the proper waste disposal channels or the required infrastructures. Bioremediation of environmental wastes has gained more attention as a cost-effective eco-friendly technology. The microbial systems are potent enough to manage most of the environmental organic and inorganic pollutants using their genetic and metabolic machinery. In this regard, the

الملخص العربي

يمثل التخلص اليومي من المخلفات في البيئة مشكلة تهدد جميع البلدان باستمرار. نظرًا للتأثيرات الضارة لهذه النفايات على البشر والحيوانات والنظام البيئي؛ تم تطوير الكثير من التقنيات للتعامل مع السمية الناتجة عن مثل هذه النفايات. على الرغم من أن المعالجات الكيميائية والفيزيائية أظهرت فعالية فيما يتعلق بمعالجة هذه الملوثات، إلا أنها تقنيات باهظة الثمن وتفتقر إلى قنوات التخلص المناسبة من النفايات أو البنى التحتية المطلوبة. لقد حظيت المعالجة الحيوية للنفايات البيئية باهتمام أكبر باعتبارها تقنية صديقة للبيئة وفعالة من حيث التكلفة. تتميز الأنظمة الميكروبية بالقوة الكافية لإدارة معظم الملوثات البيئية العضوية وغير العضوية باستخدام الآلات الوراثية والأيضية. وفي هذا الصدد، تم تناول أنواع النفايات البيئية، والتقنيات الميكروبية للمعالجة الحيوية، والمنتجات ذات القيمة

types of environmental wastes, the microbial technologies for their bioremediation, the value-added products that could microbially obtained using these waste streams in addition to other related issues have been addressed in this state of art.

Keywords: Bioremediation, Types of wastes, Genetic modification, Microbial technologies, Microbial enzymes, Value-added products.

1. Introduction

The pollution of the environment is one of the largest problems that facing the world since long times ago. It is originating from the uncontrolled discharge of the contaminants and the untreated effluents into the environment. These polluted issues are released to the environment as a result of multiple factors such as the population multiplications, the urbanization, the mining, and the industrialization (Jie et al. 2023; Ukaogo et al. 2020). The most severe concern is the co-occurrence of multiple types of pollutants in any ecosystem at the same time (Strokal et al. 2019; Yap et al. 2021). Some factors can restrict the potent remediation of multipollutants such as the incomplete quality evaluations or insufficient data collection of the contaminated sites. In such case, a comprehensive quality investigations and effective remediation technologies are intensively required in order to getting rid of these pollutants (Strokal et al. 2019).

1. المقدمة

يعد تلوث البيئة أحد أكبر المشاكل التي تواجه العالم منذ زمن طويل. وهو ينشأ من التصريف غير المنضبط للملوثات والنفايات السائلة غير المعالجة في البيئة. وتنطلق هذه القضايا الملوثة إلى البيئة نتيجة لعوامل متعددة مثل الزيادة السكانية والتحضر والتعدين والتصنيع (Jie et al. 2023; Ukaogo et al. 2020). تتمثل أكثر المخاوف خطورة في الحدوث المشترك لأنواع متعددة من الملوثات في أي نظام بيئي في نفس الوقت (Strokal et al. 2019; Yap et al. 2021). يمكن لبعض العوامل أن تحد من المعالجة الفعالة للملوثات المتعددة، مثل تقييمات الجودة غير المكتملة، أو جمع البيانات غير الكافية للمواقع الملوثة. وفي مثل هذه الحالة، هناك حاجة إلى إجراء تحقيقات نوعية شاملة وتقنيات معالجة فعالة بشكل مكثف، من أجل التخلص من هذه الملوثات (Strokal et al. 2019).

In some countries, the industrial effluents were conventionally discharged by adsorption, ion exchange, sedimentation, oxidation, electro-dialysis, and filtration (Kumar et al. 2022). However, such waste-disposal technique is challenging as in some countries it lacks the required infrastructure and/or the suitable waste disposal channels (Das et al. 2019). These traditional techniques of effluent management are expensive techniques that required high operating and capital costs (Munir et al. 2019). Recently, different microbes such as fungi, bacteria, and algae proved to effectively remediate multipollutants from the contaminated ecosystems through the enzymatic transformation of such contaminants into less-toxic and/or non-toxic substances, or even to new useful products. The process of using microbes or enzymes to getting rid of the environmental contaminants is known as bioremediation, that has gained a lot of interest and global focus research with multiple representative keywords as can be detected

في بعض البلدان، يتم تصريف النفايات السائلة الصناعية تقليديًا عن طريق الامتزاز، والتبادل الأيوني، والترسب، والأكسدة، والغسيل الكهربائي، والترشيح (Kumar et al. 2022). ومع ذلك، فإن أسلوب التخلص من النفايات هذا يمثل تحديًا، حيث أن بعض البلدان تفتقر إلى البنية التحتية المطلوبة و/أو قنوات التخلص المناسبة من النفايات (Das et al. 2019). هذه التقنيات التقليدية لإدارة النفايات السائلة هي تقنيات مكلفة تتطلب تكاليف تشغيل ورأسمالية عالية (Munir et al. 2019). وفي الآونة الأخيرة، أثبتت الكائنات الحية المختلفة مثل الفطريات والبكتيريا والطحالب أنها تعالج الملوثات المتعددة من الأنظمة البيئية الملوثة بفعالية من خلال التحول الإنزيمي لهذه الملوثات إلى مواد أقل سمية و/أو غير سامة، أو حتى إلى منتجات مفيدة جديدة. تُعرف عملية إستخدام الكائنات الحية أو الإنزيمات للتخلص من الملوثات البيئية باسم المعالجة الحيوية، والتي اكتسبت اهتمامًا كبيرًا وبحثًا عالميًا مرتكزًا على كلمات رئيسية تمثيلية متعددة يمكن اكتشافها من محرك بحث إسكوبس (الشكل 1).

from Scopus research engine (Figure 1).

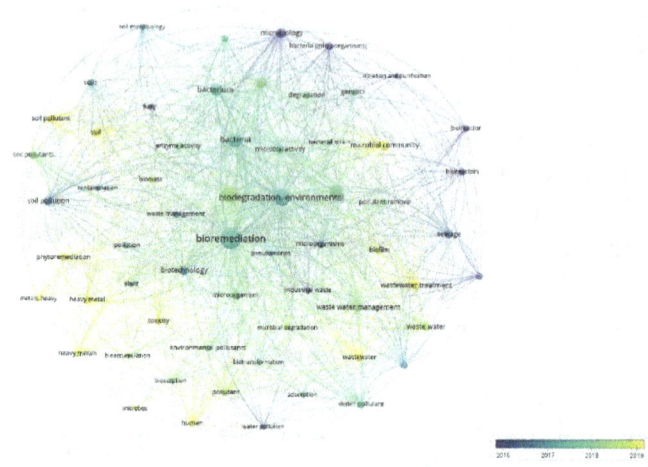

Figure 1 VOSviewer representing the world cloud keywords used in the bioremediation research topic.

الشكل 1 يمثل الكلمات الرئيسية للسحابة العالمية المستخدمة في موضوع بحث المعالجة الحيوية بإستخدام برنامج VOSviewer.

2. Definition and importance of bioremediation

2. تعريف المعالجة الحيوية وأهميتها

There has been a global increase in the applications of waste control within the sustainability activities' context. Developing countries are currently classified as those who still depend on the storage method, while the developed countries are classified as those who are able to recycle most produced wastes into new raw materials and/or to produce energy (KOÇAK and İKİZOĞLU

لقد كانت هناك زيادة عالمية في تطبيقات التحكم في النفايات ضمن سياق أنشطة الإستدامة. تُصنف الدول النامية حاليًا على أنها تلك التي لا تزال تعتمد على طريقة التخزين، بينما تُصنف الدول المتقدمة على أنها تلك التي يمكنها إعادة تدوير معظم النفايات المنتجة إلى مواد خام جديدة و/أو إنتاج الطاقة (KOÇAK and İKİZOĞLU 2020). وبالمثل، فإن

2020). Similarly, the bioremediation process is a natural one that can be an alternative to other processes such as incineration, physical removal, pollutant destruction, and catalytic degradation (Ojha et al. 2021). It is a friendless and cost-effective process that is considered an innovative and emerging technology (Singh et al. 2020b).

The bioremediation technique can be defined as the process that includes the biological degradation of environmental wastes into harmless state or the reduction of pollutants into concentrations below the limits proposed by the governing authorities (Chhaya et al. 2020). It depends on the living organisms, especially microorganisms, in order to turn the harmful chemicals and hazardous wastes into non-toxic and/or at least less toxic forms (Brusseau 2019). Although bioremediation is a green and commercially affordable technology, its efficacy can be varied from one location to another (Wang and Tam 2019). It is a low-cost technology that only needs the required organism such as bacteria, fungi, algae, or plants

عملية المعالجة الحيوية هي عملية طبيعية يمكن أن تكون بديلاً عن العمليات الأخرى مثل الحرق والإزالة المادية وتدمير الملوثات والتحلل الحفزى (Ojha et al. 2021). إنها عملية سهلة وفعالة من حيث التكلفة وتُعتبر تقنية مبتكرة وناشئة (Singh et al. 2020b).

يمكن تعريف تقنية المعالجة الحيوية على أنها العملية التي تتضمن التحلل الحيوى للنفايات البيئية إلى حالة غير ضارة أو تقليل الملوثات إلى تركيزات أقل من الحدود المقترحة من قبل السلطات الحاكمة (Chhaya et al. 2020). ويعتمد ذلك على الكائنات الحية، وخاصة الكائنات الحية الدقيقة، من أجل تحويل المواد الكيميائية الضارة والنفايات الخطرة إلى أشكال غير سامة و/أو على الأقل أقل سمية (Brusseau 2019). على الرغم من أن المعالجة الحيوية هي تقنية خضراء ومعقولة التكلفة تجاريًا، إلا أنه يمكن أن تختلف فعاليتها من موقع لآخر (Wang and Tam 2019). وهي تقنية منخفضة التكلفة لا تحتاج إلا إلى الكائن الحي المطلوب مثل البكتيريا أو الفطريات أو

to remediate and detoxify the harmful contaminants and clean them up (Zouboulis et al. 2019).

Microorganisms showed potent capacity to use the contaminants as feeding materials and as energy providers. The bioremediation process is either depends on the indigenous microbes that already exist in the contamination site for the remediation of the inhabitant pollutant, or in some cases depends on other microbes extracted from elsewhere to be added to the contaminated sites. Accordingly, the bioremediation process is hugely depending on the activities and growth of the used microbes, which in turn influenced by the environmental factors that affecting their growth and their degradation rates (Zouboulis et al. 2019). So, for effective degradation process, the exact microorganism in the exact place under optimized environmental conditions should be found (Ojha et al. 2021).

The using of microbes to convert the environmental

contaminants into safer compounds or beneficial products becomes an attractive technology, not only for the scientists, but also for businessmen (Saeed et al. 2022; El Hammoudani et al. 2021; Fierer et al. 2021). One more advantage of bioremediation is the on-site treatment of large quantities of the waste, with no need to take the wastes off to other sites. It has been reported that the bioremediation was the preferred cost effective and efficient method to remediate deepwater horizon oil spill accident, in 2010, in Mexico. All it needs was the injection of oil-degrading bacteria supplemented with some nutrients to stimulate their growth in order to get rid of 3.19 million barrels of spilled oil (Mustafa and Maryam 2023).

The bio-remediated substances might be either wastewater or solid wastes. In case of solid wastes, algae, bacteria, and fungi are used for the biodegradation of these waste in order to convert them into either less toxic or into valuable molecules that have no toxicity at all (Nriagu 2019). According to the place

مفيدة يصبح تقنية جذابة ليس فقط للعلماء، ولكن أيضًا لرجال الأعمال (Saeed et al. 2022; El Hammoudani et al. 2021; Fierer et al. 2021). وهناك ميزة أخرى للمعالجة الحيوية وهي معالجة كميات كبيرة من النفايات في الموقع، دون الحاجة إلى نقل النفايات إلى مواقع أخرى. وقد أفادت التقارير بأن المعالجة الحيوية كانت الطريقة المفضلة والفعالة من حيث التكلفة لمعالجة حادث تسرب النفط في أفق المياه العميقة، في عام 2010، في المكسيك. فكل ما تحتاج إليه هو حقن البكتيريا المحلله للنفط مع بعض العناصر الغذائية لتحفيز نموها من أجل التخلص من 3.19 مليون برميل من النفط المنسكب (Mustafa and Maryam 2023).

قد تكون المواد المعالجة حيويا إما مياه صرف أو نفايات صلبة. في حالة النفايات الصلبة، تُستخدم الطحالب والبكتيريا والفطريات للتحلل الحيوي لهذه النفايات من أجل تحويلها إما إلى جزيئات أقل سمية أو إلى جزيئات قيّمة ليس لها سمية على الإطلاق (Nriagu 2019). وفقًا لمكان العلاج، يمكن تصنيف المعالجة الحيوية

of treatment, bioremediation can be generally categories into in-situ or ex-situ bioremediation (Figure 2).

عمومًا إلى معالجة حيوية داخل الموقع أو خارج الموقع (الشكل 2).

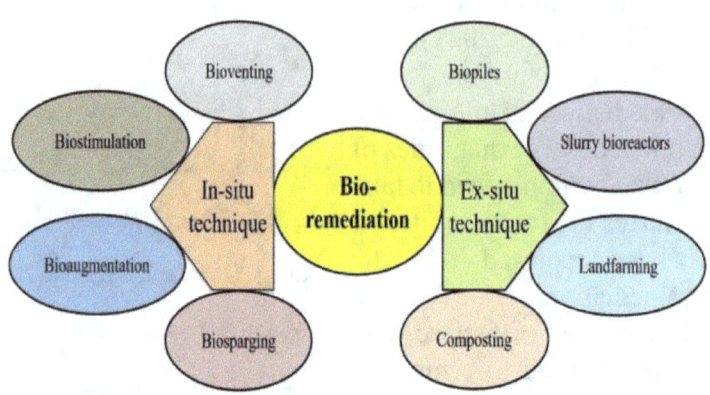

Figure 2 Types of bioremediation according to the place of treatment (Ghosh et al. 2023).

الشكل 2 أنواع المعالجة الحيوية وفقًا لمكان المعالجة (Ghosh et al. 2023).

In in-situ bioremediation, the applied microorganisms are used to treat the polluting substances at the same contaminated site, away from the ex-situ bioremediation in which the contaminated substances are transferred into other sites in order to be treated. However, in both techniques all the available biotic, abiotic, or physico-chemical conditions required for the proper growth of microorganisms such as pH, temperature, the status of the electron acceptors, etc., should

في المعالجة الحيوية داخل الموقع، تُستخدم الكائنات الحية الدقيقة المطبقة لمعالجة المواد الملوثة في نفس الموقع الملوث، بعيدًا عن المعالجة الحيوية خارج الموقع والتي يتم فيها نقل المواد الملوثة إلى مواقع أخرى من أجل معالجتها. ومع ذلك، في كلتا التقنيتين، يجب تحسين جميع الظروف الحيوية أو اللاأحيائية أو الفيزيائية والكيميائية المتاحة اللازمه للنمو الجيد للكائنات الحية الدقيقة مثل درجة الحموضة ودرجة الحرارة وحالة مستقبلات

be optimized (Ghosh et al. 2023).

2.1. In-situ bioremediation

In situ bioremediation might have some schemes or proficiencies that enhance the process. For instance, **bioventing**, in which the oxygen is directly added to the contaminated site in order to provide sufficient oxygen to the microorganisms to help them degrading the toxic pollutants such as crude oil products in polluted soil (Patel et al. 2022). Two bioventing technologies including the active or passive bioventing have been applied for the in-situ bioremediation process. In the passive bioventing, the gasses are exchanged through the ventilating wells only by the effect of the pressure resulted from the atmospheric air. While in the active bioventing, the blowers are used in order to drive the air into the ground (Zouboulis et al. 2019).

Similarly, **biosparging** is an additional oxygen injection technique in which both oxygen and nutrients are amended to the microbes in order to enhance the process of aerobic biodegradation of the existed pollutants such as petroleum and hydrocarbons (Abdulyekeen et al. 2022).

Moreover, **biostimulation** is another widely used in-situ bioremediation technique in which microbial nutrients and different electron accepting substances such as nitrogen, carbon, and oxygen are added to the polluted sites in order to enhance the production of microbial enzymes and acids (Tripathi et al. 2021).

In addition, **bioaugmentation** which is the most effective in-situ bioremediation technique has also applied. In this process, a single microbial cells or microbial consortia that have the desired catalytic capabilities are added to the polluted sites in order to enhance the overall bioremediation process (Zhang et al. 2020a). In some cases, this process includes the using of genetically modified microorganism in order to degrade the toxins and pollutants of the contaminated water or soils.

إلى الميكروبات من أجل تعزيز عملية التحلل الحيوي الهوائي للملوثات الموجودة مثل النفط والهيدروكربونات (Abdulyekeen et al. 2022).

علاوة على ذلك، فإن **التحفيز الحيوي** هو تقنية معالجة حيوية أخرى مستخدمة على نطاق واسع داخل الموقع، حيث تتم إضافة العناصر الغذائية التي تستخدمها الميكروبات والمواد المختلفة التي تقبل الإلكترون مثل النيتروجين والكربون والأكسجين، إلى المواقع الملوثة، من أجل تعزيز إنتاج الإنزيمات والأحماض الميكروبية (Tripathi et al. 2021).

بالإضافة إلى ذلك، تم أيضًا تطبيق **التعزيز الحيوي** الذي يُعد تقنية المعالجة الحيوية الأكثر فعالية داخل الموقع. في هذه العملية، تتم إضافة خلايا ميكروبية واحدة أو إتحادات ميكروبية تتمتع بالقدرات التحفيزية المطلوبة إلى المواقع الملوثة، من أجل تعزيز عملية المعالجة الحيوية الشاملة (Zhang et al. 2020a). وفي بعض الحالات، تشمل هذه العملية استخدام الكائنات الحية الدقيقة المعدلة وراثيًا، من أجل تحليل السموم والملوثات الموجوده في المياه أو التربة الملوثة.

2.2. Ex-situ bioremediation

On the other hand, in terms of the ex-situ bioremediation, **land farming** is straight forward low-cost process that requires less operational equipment that has been predominantly applied for the bioremediation of hydrocarbon-contaminated soil, pit sludge, and oily sludge. The polluted material naturally has microbial strains that can convert the pollutant substances into carbon dioxide, water, and microbial biomass. It has been reported that, if the pollutants located at >1.7 m of the contaminated soil, it should be excavated. While if they located at <1 m, no excavation is needed (Ghosh et al. 2023). Land farming process is depending on the placing of the contaminated waste materials onto lined beds submitted for aeration with simultaneous optimization of some conditions such as the soil pH, the moisture content, the nutrients, and the amount of oxygen. However, it is just limited to treat only 10-30 cm of contaminated soil.

2.2. المعالجة الحيوية خارج الموقع

ومن ناحية أخرى، فيما يتعلق بالمعالجة الحيوية خارج الموقع، فإن **زراعة الأراضي** هي عملية بسيطة منخفضة التكلفة، تتطلب معدات تشغيلية أقل، تم تطبيقها في الغالب على المعالجة الحيوية للتربة الملوثة بالهيدروكربونات وحمأة الحفرة والحمأة النفطية. تحتوي المادة الملوثة بشكل طبيعي على سلالات ميكروبية يمكنها تحويل المواد الملوثة إلى ثاني أكسيد الكربون والماء والكتلة الحيوية الميكروبية. وقد تم الإبلاغ عنه إذا كانت الملوثات الموجودة على مسافة تزيد عن 1.7 متر من التربة الملوثة، فيجب دفنها. بينما إذا كانت تقع على مسافة 1 متر، فلا يلزم إجراء أي دفن (Ghosh et al. 2023). تعتمد عملية زراعة الأراضي على وضع مواد النفايات الملوثة على أحواض مبطنة، يتم تقديمها للتهوية مع تحسين متزامن لبعض الظروف مثل درجة حموضة التربة، ومحتوى الرطوبة، والعناصر الغذائية، وكمية الأكسجين. ومع ذلك، يقتصر الأمر على معالجة 10-30 سم فقط من التربة الملوثة.

On the other hand, **biopiles** are a combination of composting and farming, where the polluted materials are transported after their excavation to other places that can provide aeration and systems for leachate collection, and help for the bioremediation process. It can also support the growth and activity of both aerobic and anaerobic microbes. The biopiles existed microbes are producing heat during their biodegradation activity that elevating the temperature to be in the range of 55-65°C. Biopiles are usually covered with plastic-based covers in order to prevent the evaporation or the volatilization of the components that have low molecular-weights (Zheng et al. 2021).

Moreover, the **composting** process is a sustainable one that resulted in valuable products. It combines the safe organic compounds such as animal manure, vegetative wastes, wood chips, and hay that can be microbially decomposed and resulted in increasing the soil porosity in addition to increasing the microbial population. Three

ومن ناحية أخرى، تُعد **الكتل الحيوية** مزيجًا من التسميد والزراعة، حيث يتم نقل المواد الملوثة بعد التنقيب عنها إلى أماكن أخرى يمكنها توفير التهوية وأنظمة جمع العصارة، والمساعدة في عملية المعالجة الحيوية. كما يمكن أن يدعم نمو ونشاط كل من الميكروبات الهوائية واللاهوائية. تنتج الميكروبات الموجودة في الأكوام الحيوية الحرارة أثناء نشاط التحلل الحيوى الذي يرفع درجة الحرارة في نطاق 55-65 درجة مئوية. عادةً ما تتم تغطية الأكوام الحيوية بأغطية بلاستيكية لمنع تبخر أو تطاير المكونات ذات الأوزان الجزيئية المنخفضة (Zheng et al. 2021).

علاوة على ذلك، فإن عملية **التحويل إلى سماد** هي عملية مستدامة، تؤدي إلى منتجات قيمة. فهي تجمع بين المركبات العضوية الآمنة مثل السماد الحيواني المخلفات النباتية ورقائق الخشب والقش، التي يمكن أن تتحلل ميكروبيًا وينتج عنها زيادة مسامية التربة، بالإضافة إلى زيادة التجمعات الميكروبية. تم تطبيق ثلاثة أنواع

composting types have been applied which including: static aerated piles or mechanically agitated vessels, or long/narrow piles with mobile equipment. All the methods prefer the temperature range of 54-65°C (Golbaz et al. 2021).

In addition, **slurry bioreactors** are considered to be the most recent and powerful technique of the ex-situ bioremediation technologies. It involves the microbial degradation of pollutants existed in soil, groundwater, and wastewater. The polluted substances are transferred into an aeriated chamber where the organic compounds are microbially degraded and forming sludge (Hussain et al. 2022). Generally, slurry bioreactor is a vessel in which the bioremediation of soil or aqueous pollutants is increased through the creation of gas/liquid/solid mixing phase. The working operations might be batch, sequencing batch, fed batch, or continuous operations, that provide controllable conditions of pH, temperature, aeration as natural environment for the microbes to enhance their growth. The slurry bioreactor

من التسميد والتي تشمل: أكوام هوائية ثابتة أو أوعية متحركة ميكانيكيًا أو أكوام طويلة/ضيقة مزودة بمعدات متحركة. وتفضل جميع الطرق نطاق درجة الحرارة من 54 إلى 65 درجة مئوية (Golbaz et al. 2021).

بالإضافة إلى ذلك، تعتبر **المفاعلات الحيوية الطينية** أحدث وأقوى تقنية لتقنيات المعالجة الحيوية خارج الموقع. ويتضمن التحلل الميكروبي للملوثات الموجودة في التربة والمياه الجوفية ومياه الصرف. حيث يتم نقل المواد الملوثة إلى غرفة هوائية، حيث تتحلل المركبات العضوية ميكروبيًا وتشكل حمأة (Hussain et al. 2022). بشكل عام، فإن المفاعل الحيوي الطيني هو وعاء تتم فيه زيادة المعالجة الحيوية للتربة أو الملوثات المائية، من خلال إنشاء مرحلة خلط الغاز/السائل/الصلب. قد تكون عمليات العمل عبارة عن دفعة واحدة أو دفعة متسلسله أو دفعة مغذية أو عمليات مستمرة، توفر ظروفًا يمكن التحكم فيها من حيث درجة الحموضة ودرجة الحرارة والتهوية كبيئة طبيعية للميكروبات لتعزيز نموها. والمفاعل الحيوي الطيني هو عملية المعالجة الحيوية للنفايات الصلبة الأكثر

is the most advanced solid waste bioremediation process that is not time-consuming one and can efficiently and quickly deals with most household, industrial, agricultural, or municipal solid wastes as feed-stocks for the production of value-added products, biochemical, and bioenergy (Figure 3). The produced bioenergy such as biofuels are considered the 2nd-generation biofuels compared with the 1st-generation ones that include food wastes such as barely, wheat, corn, sugarcane etc., (Ghosh et al. 2023).

تقدمًا، والتي لا تستغرق وقتًا طويلاً، ويمكن أن تتعامل بكفاءة وسرعة مع معظم النفايات الصلبة المنزلية أو الصناعية أو الزراعية أو البلدية، كمواد خام لإنتاج المنتجات ذات قيمة قوية والمواد الكيميائية الحيوية والطاقة الحيوية (الشكل 3). تعتبر الطاقة الحيوية المنتجة مثل الوقود الحيوي هي الجيل الثاني من الوقود الحيوي مقارنةً بالجيل الأول الذي يتضمن مخلفات الطعام مثل الشعير، والقمح، والذرة، وقصب السكر وما إلى ذلك، (Ghosh et al. 2023).

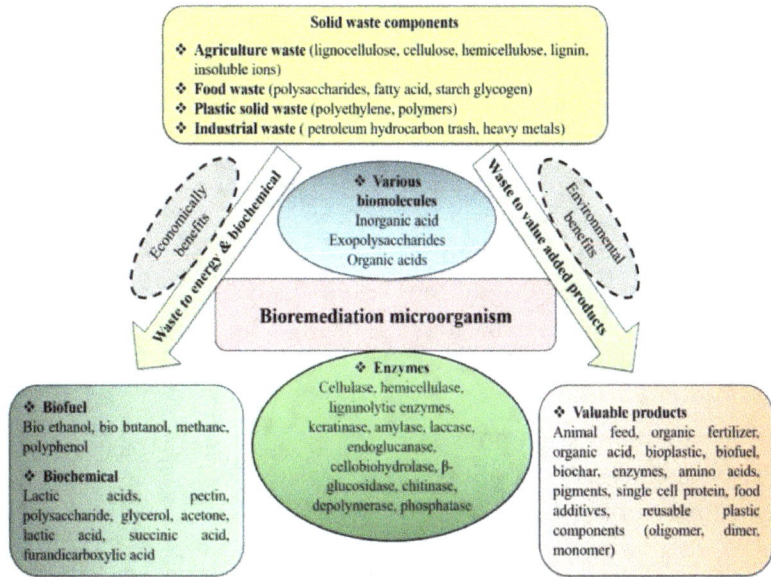

Figure 3 The microbial conversion of solid wastes into value-added products (Ghosh et al. 2023; Mohanan et al. 2020; Zabermawi et al. 2022a; Ghorai and Ghosh 2022a, b; Chilakamarry et al. 2022; Adegboye et al. 2021).

الشكل 3 التحويل الميكروبي للنفايات الصلبة إلى منتجات ذات قيمة مضافة (Ghosh et al. 2023; Mohanan et al. 2020; Zabermawi et al. 2022a; Ghorai and Ghosh 2022a, b; Chilakamarry et al. 2022; Adegboye et al. 2021).

3. Microbial Technologies in Bioremediation

Multiple mechanisms have been detected for the ability of microbes to remove pollutants from the contaminated sites. Two main and big categories namely mobilization and immobilization can precisely describe these mechanisms (Verma and Kuila 2019).

3. التقنيات الميكروبية في المعالجة الحيوية

تم الكشف عن آليات متعددة لقدرة الميكروبات على إزالة الملوثات من المواقع الملوثة. يمكن لفئتين رئيسيتين كبيرتين، وهما التعبئة والتثبيت، أن تصف هذه الآليات بدقة (Verma and Kuila 2019). وتنتمي بعض التقنيات، بما في

Some techniques including bioleaching, enzymatic oxidation, enzymatic reduction, bioaugmentation, and biostimulation are belonging to the mobilization process. While complexation, bioaccumulation, biosorption, solidification by precipitation processes are belonging to the immobilization technology (Ayangbenro et al. 2019).

In the immobilization process, microbes convert the pollutant into other forms that will not be available in the environment, such as the bioconversion of the NH_3 nitrogen into organic nitrogen. This process is the best applied one in the heavy metals' bioremediation in high concentrations. During the immobilization process, the contaminated soils can be transferred to another location in order to be submitted for Chelating the polluting heavy metals by the action of microbes, in a process known as ex-situ bioremediation. While, in other cases, the term of in-situ bioremediation is used, where the polluting material is treated at the same contaminated site. For instance, heavy metals have been immobilized by some

ذلك الترشيح الحيوي، والأكسدة الإنزيمية، والإختزال الإنزيمي، والزيادة الحيوية، والتحفيز الحيوي، إلى عملية التعبئة. فى حين أن التعقيد والتراكم الحيوي، والامتصاص الحيوي، والتصلب عن طريق عمليات الترسيب، إلى تقنية التثبيت (Ayangbenro et al. 2019).

في عملية التثبيت، تقوم الميكروبات بتحويل الملوثات إلى أشكال أخرى لن تكون متاحة في البيئة، مثل التحويل الحيوي من نيتروجين إلي نيتروجين عضوي. هذه العملية هي أفضل عملية يتم تطبيقها في المعالجة الحيوية للمعادن الثقيلة بتركيزات عالية. أثناء عملية التثبيت، يمكن نقل التربة الملوثة إلى موقع آخر من أجل تحلل المعادن الثقيلة بفعل الميكروبات، في عملية تُعرف باسم المعالجة الحيوية خارج الموقع. وفي حالات أخرى، يُستخدم مصطلح المعالجة الحيوية في الموقع، حيث تتم معالجة المادة الملوثة في نفس الموقع الملوث. على سبيل المثال، تم تثبيت المعادن الثقيلة بواسطة بعض البكتيريا مثل *E. asburiae B. cereus* (Fashola et al. 2020). وتجدر الإشارة إلى أنه يمكن

bacteria such as E. asburiae and B. cereus (Fashola et al. 2020). It worth mentioning that, the microbes can be protected from toxic compounds in the contaminated sites by the formation of hydrophobic or solvent efflux pump that prevents their outer cell membranes from damage (Verma and Kuila 2019).

حماية الميكروبات من المركبات السامة داخل المواقع الملوثة عن طريق تكوين مضخة تدفق مذيبة أو كارهة للماء، والتى تمنع أغشية الخلايا الخارجية من التلف (Verma and Kuila 2019).

3.1. Enzymatic oxidation
3.2. Enzymatic reduction

In this process the microbes are using enzymes that have an opposite action to the enzymatic oxidation process. In this case, the contaminant is converted into a reduced state that converts it into insoluble molecule. This process is usually adopted by the obligate and facultative anaerobes. For instance, chrome reductase enzymes are reducing the hexavalent chromium ions into trivalent ions in order to reduce its toxicity. In addition, the azoreductase enzymes have been used for the reduction of azo compounds through the azo bonds cleavage (Saxena et al. 2020).

3.1. الأكسدة الإنزيمية

3.2.الإختزال الإنزيمى

في هذه العملية تستخدم الميكروبات إنزيمات لها عمل معاكس لعملية الأكسدة الإنزيمية. وفي هذه الحالة، يتم تحويل الملوثات إلى حالة مختزلة، تحولها إلى جزيء غير قابل للذوبان. عادة ما يتم اعتماد هذه العملية من قبل اللاهوائيات الإجبارية والاختيارية. على سبيل المثال، تعمل إنزيمات مختزلة الكروم على تقليل أيونات الكروم سداسية التكافؤ، إلى أيونات ثلاثية التكافؤ، من أجل تقليل سميتها. بالإضافة إلى ذلك، تم استخدام إنزيمات "الأزورديكتز" لتقليل مركبات الأزو من خلال شق روابط الأزو (Saxena et al. 2020).

3.3. Bioaugmentation

It is a process where external microorganisms are added to the polluted sites in order to augment the existed microbes and to feed on the toxic pollutants and remediate them from these sites. This process is a rapid, cost-effective, potent bioremediation process (Mahmoud 2021). However, in other bioaugmentation processes, the resident microbes in the polluted sites are isolated and subjected for genetic modifications and subsequently returned to the same site in order to enhance their remediation capability, as these naturally existed microbes might not be able to remediate the existed pollutants. In case of adding external microbes to promote the degradation of a pollutant in a specific site; the effectiveness of the bioremediation process is extremely depending on the ability of these microbes to be in harmony with the resident microbes and also their ability to be adapted to the new habitats, as well (Babalola et al. 2019). In the same context, the bioremediation of

3.3. التعزيز الحيوي

وهي عملية يتم فيها إضافة الكائنات الحية الدقيقة الخارجية إلى المواقع الملوثة من أجل زيادة الميكروبات الموجودة والتغذية بالملوثات السامة ومعالجتها من هذه المواقع. وهذه العملية عبارة عن عملية معالجة حيوية سريعة وقوية وفعالة من حيث التكلفة (Mahmoud 2021). ومع ذلك، في عمليات التعزيز الحيوى الأخرى، يتم عزل الميكروبات الموجودة في المواقع الملوثة وإخضاعها للتعديلات الجينية، ثم إعادتها إلى نفس الموقع من أجل تعزيز قدرتها على المعالجة، حيث قد لا تتمكن هذه الميكروبات الموجودة بشكل طبيعي من معالجة الملوثات الموجودة. في حالة إضافة ميكروبات خارجية لتعزيز تحلل الملوثات في موقع معين؛ تعتمد فعالية عملية المعالجة الحيوية بشكل كبير على قدرة هذه الميكروبات على الانسجام مع الميكروبات الموجودة، وكذلك قدرتها على التكيف مع الموائل الجديدة أيضا (Babalola et al. 2019). وفي نفس السياق، كانت المعالجة الحيوية لمركبات "النتروفينول" الموجودة في التربة الملوثة

nitrophenolic compounds existed in a pesticides contaminated soil to less toxic forms was achieved through the augmentation of *Bukholderia* sp. FDS-1 at 30°C and slightly acidic pH (Ojuederie et al. 2021). Other bioaugmentation experiments can be found in Table 1.

بالمبيدات الحشرية إلى أشكال أقل سمية، من خلال زيادة *Bukholderia* sp. FDS-1 عند 30 درجة مئوية ودرجة حموضة حمضية قليلة (Ojuederie et al. 2021). يمكن العثور على تجارب التعزيز الحيوي الأخرى في الجدول 1.

Table 1 Studies of different bioremediation mechanisms (Ayilara and Babalola 2023).

الجدول 1 دراسات آليات المعالجة الحيوية المختلفة (Ayilara and Babalola 2023).

The bioremediation mechanism آلية المعالجة الحيوية	The pollutant الملوث	Microorganism الكائن الحي الدقيق	References المراجع
Bioimmobilization التثبيت الحيوي	Lead الرصاص	*Bacillus subtilis*	Qiao et al., 2019
Bioattenuation التوهين الحيوي	Copper and lead النحاس والرصاص	*Pseudomonas sp.*	Nanda et al., 2019
Bioprecipitation الترسيب الحيوي	Cadmium الكادميوم	*Cupriavidus sp.*	Li et al., 2019
Bioaugmentation التعزيز الحيوي	Aluminium, lead, cadmium, and copper الألمونيوم والرصاص والكادميوم والنحاس	*Rhodococcus sp., Lysinibacillus sp.* and *Bacillus sp.,*	Nanda et al., 2019
Enzymatic reduction الإختزال الإنزيمي	Congo red dye	*Oudemansiella canarii*	Iark et al., 2019

	صبغة الكونجو الحمراء		
Biosorption الامتصاص الحيوي	Nickel النيكل	*Bacillus sp.*	Taran et al., 2019
Enzymatic reductase الإنزيم الإختزالى	Azo dyes أصباغ الأزو	*Lysinibacillus sphaericus*	Lu et al., 2020
Intracellular sequestration العزل داخل الخلايا	Copper النحاس	*Sulfolobus solfataricus*	Thakare et al., 2021
Extracellular sequestration العزل خارج الخلايا	Copper, zinc and cadmium النحاس والزنك والكادميوم	*Desulfovibrio desulfuricans*	Thakare et al., 2021

3.4. Biostimulation

It is the process where the nutrients such as potassium, phosphorus, and nitrogen, or other metabolites, or enzymes, or biosurfactants, or electron acceptors/donors are added to the contaminated sites in order to improve the activity of the indigenous microbes and elevate their remediation capabilities. It is an effective process that is affordable and eco-friendly. When comparing bioaugmentation and biostimulation process, the later one is preferred as the resident microbes are more competitive than the externally added ones, beside the preservation of the natural microbial balance and diversity in the treated sites (Sayed et al. 2021). The

3.4. التحفيز الحيوي

وهي العملية التي تتم فيها إضافة العناصر الغذائية مثل البوتاسيوم والفوسفور والنيتروجين، أو المستقلبات الأخرى، أو الإنزيمات، أو الموادالحيوية السطحية، أو مستقبلات/مانحات الإلكترونات، إلى المواقع الملوثة، من أجل تحسين نشاط الميكروبات الأصلية وزيادة قدراتها العلاجية. إنها عملية فعالة ميسورة التكلفة وصديقة للبيئة. عند مقارنة التعزيز الحيوى و التحفيز الحيوي، يتم تفضيل الأحدث لأن الميكروبات المقيمة أكثر قدرة على المنافسة من الميكروبات المضافة خارجيًا، إلى جانب الحفاظ على التوازن الميكروبي الطبيعي والتنوع داخل المواقع المعالجة (Sayed et al. 2021). أظهرت عملية

biostimulation process has showed a potent activity in the bioremediation of heavy metals using different microbial genera such as *Pseudomonas* sp., *Klebsiella* sp., *Bacillus* sp., *Staphylococcus* sp., *Rhodococcus* sp., and *Citrobacter* sp. (Nivetha et al. 2023). However, some environmental harms may occur due to the biostimulation process such as eutrophication that might result from the addition of excess nutrients. Similarly, the addition of synthetic nutrients or chemicals might be another source of environmental pollution that is completely defeating the purpose of the bioremediation process (Ayilara and Babalola 2023).

3.5. Bioleaching

In this process, the acidophilic microbes are used to solubilize the heavy metals that are existed in a solid form of the sediments. This process is hugely beneficial for the liberation of iron and sulfur, from the iron and sulfur pollutants (Sun et al. 2021; Bhandari et al. 2023). This mission is usually achieved using the iron- or sulfur-oxidizing bacteria that able to change the medium pH to be acidic, and hence solubilizing the already immobilized heavy metals to be released into the aqueous solution such as: *Penicillium* sp., *Aspergillus* sp., *Rhizopus* sp., *Mucor* sp., *Cladosporium* sp., and *A. thiooxidans* (Medfu Tarekegn et al. 2020).

3.6. Biosorption

This is a bioremediation process where the heavy metals are adsorbed from pollutants through chelation, complexation, ion displacement, and physical interaction with electrostatic forces (Mahmoud 2021). The metals are generally linked to the active groups of the chemical compounds existed in the cells surface layers of bacteria, fungi, or algae, and resulted in the movement of the ions between the negatively charged active groups of the outer parts of the microbial cells and the metal cations. Multiple microbial genera have been reported as capable or using the biosorption process through their bioremediation of pollutants such as *Streptomyces* sp., *Bacillus* sp.,

والتي يمكنها تغيير درجة الحموضة المتوسطة لتصبح حمضية، ومن ثم إذابة المعادن الثقيلة المثبتة بالفعل ليتم إطلاقها في المحلول المائي مثل: *Penicillium* sp., *Aspergillus* sp., *Rhizopus* sp., *Mucor* sp., *Cladosporium* sp.، و *A. thiooxidans* (Medfu Tarekegn et al. 2020).

6.3.الإمتصاص الحيوي

وهي عملية معالجة حيوية يتم فيها امتصاص المعادن الثقيلة من الملوثات، عن طريق الاستخلاب والتعقيد وإزاحة الأيونات والتفاعل الفيزيائي مع القوى الإلكتروستاتيكية (Mahmoud 2021). ترتبط المعادن بشكل عام بالمجموعات النشطة للمركبات الكيميائية الموجودة في الطبقات السطحية للخلايا من البكتيريا، أو الفطريات، أو الطحالب وتؤدي إلى حركة الأيونات بين المجموعات النشطة سالبة الشحنة للأجزاء الخارجية من الخلايا الميكروبية والكاتيونات المعدنية. تم استخدام أجناس ميكروبية متعددة على أنها قادرة على استخدام عملية الامتصاص الحيوي من خلال معالجتها البيولوجية للملوثات مثل *Streptomyces* sp.

and *Rhodococcus* sp. (Baltazar et al. 2019; Sedlakova-Kadukova et al. 2019), while complexation strategy depending on the using of ligand that able to form a complex with the inorganic metals in the polluted sites. The ligands are either high molecular molecules, or siderophores, or low-molecular weight organic acids such as alcohols, citric acids, or tricarboxylic acids. The complexation process is almost occurring when the surface polymeric substances of the microbes are interacted with the heavy metals of the polluted sites (Xie et al. 2020).

Multiple bacteria such as *B. lichenformis* and *Rhodobacter blasticus* have been reported to bind the heavy metals in the polluted environments through complexation process (Wang et al. 2019; Bai et al. 2019). When the microbes existed in heavy metal polluted environments, and simultaneously suffered from iron-deficiency, they are producing siderophores such as phenolates, or catecholates that able to make complexes with heavy metals and resulted in increasing their solubility. Siderophores are also able to produce reactive oxygen species that can remediate the organic contaminants and resulted in enhanced bioremediation process (Albelda-Berenguer et al. 2019). Moreover, scientists discovered some bacterial species such as Comamonas sp., that has a potential capacity to flourish in contaminated habitats and strong adaptation to several environmental conditions. These bacteria have showed remarkable ability to detoxify heavy metals from different contaminated sites using several mechanisms like: biosorption, redox transformation, and intracellular sequestration (Hussein et al. 2023).

3.7. Bioaccumulation

The bioaccumulation process is the one in which the rate of the absorption of certain compound is higher than the rate of its lost. It is resulted in the accumulation of the compounds in the intra-cellular part of the microbial cells (Sharma et al. 2022). Various mechanisms such as ion pumps, protein channel, and carrier- mediated transport are known for the ability of heavy metals to cross the outer membranes of the microbial cells. Different microbes have been reported as heavy metals bio accumulating microbes. For instance, *Pseudomonas putida* has been reported as cadmium bio remediating bacteria, while *Rhizopus arrhizus* has been reported for remediating mercury, and *Aspergillus niger* for remediating thorium element (Sharma et al. 2022).

3.8. Precipitation

This process includes the conversion of pollutants or heavy metals into crystals or precipitates in order to reduce their toxicity through enzymatic activity or galactosis process of the secondary metabolites (Sharma et al. 2022). For instance, the sulfate-reducing-bacteria are able to convert organo-phosphate into ortho-phosphate at alkaline conditions. Similarly, both of *Oceanobacillus indicireducens* and *Bacillus subtilis* showed specific activity to precipitate heavy

وقناة البروتين والنقل بواسطة الناقل بقدرة المعادن الثقيلة على عبور الأغشية الخارجية للخلايا الميكروبية. هناك ميكروبات مختلفة تعمل على تجمع المعادن الثقيلة بيولوجيًا. على سبيل المثال، *Pseudomonas putida* تستخدم فى المعالجة الحيوية للكادميوم، بينما *Rhizopus arrhizus* فى المعالجة الحيوية للزئبق، و *Aspergillus niger* فى المعالجة الحيوية لعنصر الثوريوم (Sharma et al. 2022).

3.8.الترسيب

تتضمن هذه العملية تحويل الملوثات أو المعادن الثقيلة، إلى بلورات أو رواسب من أجل تقليل سميتها، من خلال النشاط الإنزيمي أو عملية الجلاكسو للمركبات الايضية الثانوية (Sharma et al. 2022). على سبيل المثال، البكتيريا المختزلة للكبريتات قادرة على تحويل الفوسفات العضوي إلى أورثو فوسفات في الظروف القلوية. وبالمثل، أظهر كل من *Oceanobacillus indicireducens* و *Bacillus subtilis* نشاطًا محددًا لترسب المعادن الثقيلة في البيئات الملوثة

metals in the polluted environments (Maity et al. 2019). Generally, most applied bioremediation processes of the solid waste are summarized in Figure 4.

(Maity et al. 2019). وبشكل عام، يتم تلخيص معظم عمليات المعالجة الحيوية المطبقة للنفايات الصلبة في الشكل 4.

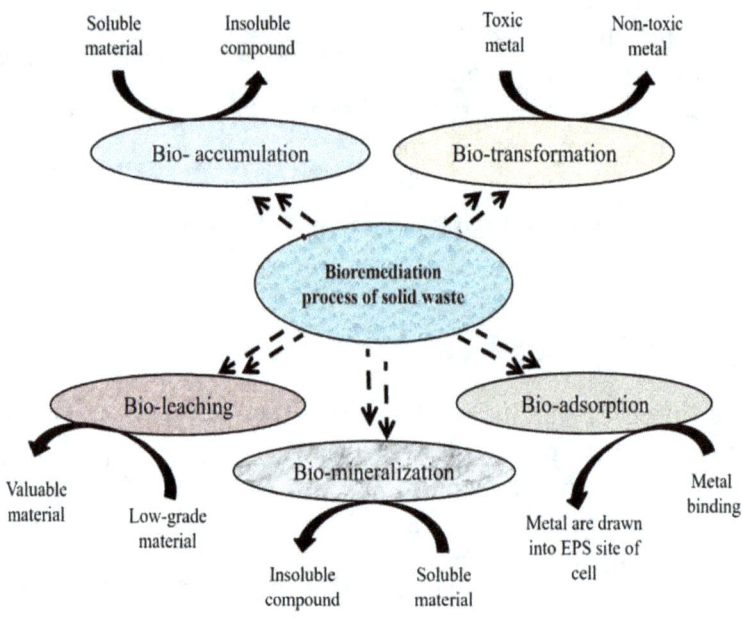

Figure 4 Summarization of the most applied bioremediation technologies of solid wastes (Sayara et al. 2020; Kapahi and Sachdeva 2019; Verma et al. 2021; Dabe et al. 2019; Ghosh et al. 2023).

الشكل 4 تلخيص تقنيات المعالجة الحيوية الأكثر تطبيقًا للنفايات الصلبة (Sayara et al. 2020; Kapahi and Sachdeva 2019; Verma et al. 2021; Dabe et al. 2019; Ghosh et al. 2023).

4. Using of microbial enzymes for multipollutant bioremediation

According to the type of pollutants, they can be categorized into non-biodegradable or biodegradable. The first category is known as those that can be decayed by microbial enzymes into non-toxic simpler compounds, such as kitchen wastes, farm by products, cow-manure, and wood. While the second category represents those pollutants that are not naturally broken down into harmless compounds such as some insecticides, pesticides, synthetic plastics, heavy metals, and demolition residues (Mangla et al. 2019). The difference between non-biodegradable and biodegradable pollutants can be shown in Figure 5.

4. إستخدام الإنزيمات الميكروبية للمعالجة الحيوية للملوثات المتعددة

وفقًا لنوع الملوثات، يمكن تصنيفها إلى ملوثات قابلة للتحلل، أو ملوثات غير قابلة للتحلل. تُعرف الملوثات القابلة للتحلل بأنها تلك التي يمكن أن تتحلل بفعل الإنزيمات الميكروبية إلى مركبات أبسط غير سامة مثل مخلفات المطبخ، والزراعة والخشب، في حين أن الملوثات الغير قابلة للتحلل هي تلك الملوثات التي لا تتحلل بشكل طبيعي إلى مركبات غير ضارة مثل بعض المبيدات الحشرية، والمواد البلاستيكية الاصطناعية، والمعادن الثقيلة، ومخلفات الهدم (Mangla et al. 2019). يمكن توضيح الفرق بين الملوثات الغير قابلة للتحلل الحيوي والملوثات القابلة للتحلل الحيوي في الشكل 5.

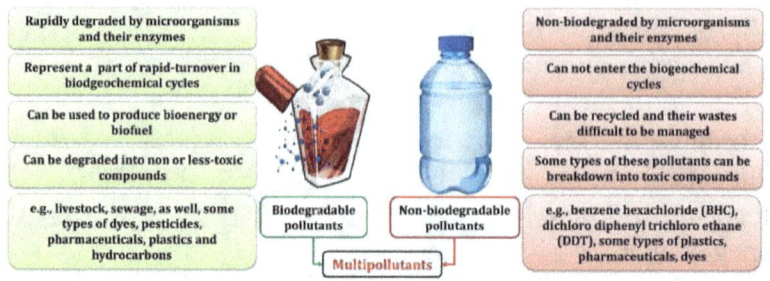

Figure 5 Characteristics of non-biodegradable and biodegradable contaminants in the environment (Narayanan et al. 2023).

الشكل 5 خصائص الملوثات الغير قابلة للتحلل الحيوى والقابلة للتحلل في البيئة (Narayanan et al. 2023).

Multiple microbial enzymes such as laccases, proteases, lipases, dehydrogenases, etc., have been showed potent activities towards the degradation or the transformation of multipollutants such as aromatic hydrocarbons, polymers, dyes, detergents, pesticides, etc., (Bhandari et al. 2021). A brief description of the identity and action of some of these enzymes will be discussed as follows:

أظهرت الإنزيمات الميكروبية المتعددة مثل اللاكياز، والبروتياز، والليباز، والديهيدروجينياز، وما إلى ذلك، أنشطة فعالة تجاه تحلل أو تحويل الملوثات المتعددة مثل الهيدروكربونات العطرية، والبوليمرات، والأصباغ، والمنظفات، والمبيدات الحشرية، وما إلى ذلك (Bhandari et al. 2021)، وسيتم مناقشة وصف موجز لهوية وفاعلية بعض هذه الإنزيمات على النحو التالي:

4.1. Hydrolytic enzymes

The only pollutants that can transfer via cell structures are those with molecular

4.1. الإنزيمات المحللة

الملوثات الوحيدة التي يمكن أن تنتقل عبر الهياكل الخلوية هي تلك التي يقل وزنها

weight of less than 600 kg/mol, which clearly states that the activity of extra-cellular enzymes of the deterioration of such pollutants is critical (Dong and Zhang 2021). These hydrolytic enzymes can effectively break down the chemical bonds of the harmful pollutants to be fewer toxic compounds. The enzymes are perfectly degrading pollutants such as carbamate insecticides, organophosphates, and oil spills. They have a lot of benefits such as the easy accessibility to the pollutants, they do not need co-factor stereoselectivity, and their tolerance to the added water-miscible solvents (Gangola et al. 2019). Hydrolases are almost catalyzing the initial step in the hydrolysis of ester-bond to quickly biodegrading the pollutants. They have a wide range of applications in food, biochemical, and chemical industries. Amylases, xylanases, lipases, and proteases, are examples of the hydrolytic enzymes (Thapa et al. 2019).

الجزيئي عن 600 كجم/مول، مما ينص بوضوح على أن نشاط الإنزيمات خارج الخلايا لتدهور هذه الملوثات أمر بالغ الأهمية (Dong and Zhang 2021). ويمكن لهذه الإنزيمات التحليلية المائية أن تكسر الروابط الكيميائية للملوثات الضارة لتكون أقل سمية. تعمل الإنزيمات على تحلل الملوثات تمامًا مثل المبيدات الحشرية الكاربامينية، والفوسفات العضوية وتسربات النفط. ولها الكثير من الفوائد مثل سهولة الوصول إلى الملوثات، ولا تحتاج إلى الانتقائية الفراغية للعوامل المشتركة، وتحملها للمذيبات المضافة القابلة للامتزاج المائي (Gangola et al. 2019). تعمل الهيدروليزات تقريبًا على تحفيز الخطوة الأولى في التحلل المائي لرابطة الإستر لتحلل الملوثات بيولوجيًا بسرعة. لديهم مجموعة واسعة من التطبيقات في الصناعات الغذائية والكيميائية الحيوية والكيميائية. تعد الأميليز، والزيلانيز، والليبيز، والبروتيز، أمثلة على الإنزيمات المحللة (Thapa et al. 2019).

4.2. Lipases

Lipases are a group of enzymes that are produced by many microbes, and able to break down the oil and lipids into their initial chemical structures. They showed high reactivity to reduce the hydrocarbon content in oily polluted sites (Yusoff et al. 2020). They are catalyzing the hydrolysis of the tri-acyglycerols into fatty acids and glycerols, in addition to their ability to catalyze other reactions such as esterification, alcoholysis, and the aminolysis (Chandra et al. 2020). They have been proved to be produced by yeast strains such as *Candida rugosa* that showed an ability to hydrolyze triolein in oil-water tanks into diloein, monoolein, and glycerol through the breakdown of the ester linkages. Generally, the microbial lipase activities were accounted as the valuable marker characteristics for the evaluation of hydrocarbons deterioration in contaminated soils (Narayanan et al. 2023).

4.2. الليبيزيز

إنزيمات الليبيازيز هي مجموعة من الإنزيمات التي تنتجها العديد من الميكروبات، وهي قادرة على تحليل الزيوت والدهون إلى بنياتها الكيميائية الأولية. وقد أظهرت تفاعلاً عالياً لتقليل محتوى الهيدروكربونات في المواقع النفطية الملوثة بالزيت (Yusoff et al. 2020). وهي تحفز التحليل المائي لثلاثي الجلسرين إلى أحماض دهنية وجلسرين، بالإضافة إلى قدرتها على تحفيز التفاعلات الأخرى مثل الإسترة، وتحلل الكحول، وتحلل الأمينيات (.Chandra et al 2020). وقد ثبت أنه يتم إنتاجها عن طريق سلالات الخميرة مثل *Candida rugosa* التي أظهرت القدرة على تحليل ثلاثي الأوليين في خزانات النفط والمياه إلى ديلوين ومونولين وجلسرين، من خلال تحلل روابط الإستر. وبصفة عامة، تم احتساب أنشطة الليباز الميكروبي باعتبارها الخصائص المميزة لتقييم تدهور الهيدروكربونات في التربة الملوثة (Narayanan et al. 2023).

4.3. Oxidoreductases

Large numbers of bacteria and fungi are able to detoxify the structure of the organic pollutants through the oxidoreductases enzymes that act on the oxidative coupling interactions (Espinosa-Ortiz et al. 2022).

These enzymes facilitating the transfer of electrons resulted from the cleaving of chemical bonds from one organic pollutant to other chemical molecules which resulted in the obtaining of energy by the enzyme-producing microbes (Koutra et al. 2021). Oxidoreductases have been involved in the detoxification of some pollutants such as polyphenols and anilinic substances through their polymerization and/or co-polymerization with other chemical compounds or through their stick to humic acids (Zhou et al. 2022). Multiple detoxification processes of various compounds or elements such as azo dyes and irradiated metals were performed through the oxidation/reduction interactions (Al-Tohamy et al. 2020; Al-Tohamy et al. 2021;

3.4. إنزيمات الأكسدة والاختزال

تستطيع أعداد كبيرة من البكتيريا والفطريات إزالة السموم من بنية الملوثات العضوية، من خلال إنزيمات الأكسدة والاختزال التي تعمل على تفاعلات الاقتران المؤكسدة (Espinosa-Ortiz et al. 2022).

تعمل هذه الإنزيمات على تسهيل نقل الإلكترونات الناتجة عن انقسام الروابط الكيميائية من أحد الملوثات العضوية إلى جزيئات كيميائية أخرى مما يؤدي إلى الحصول على الطاقة عن طريق الميكروبات المنتجة للإنزيمات (Koutra et al. 2021). شاركت إنزيمات الأكسدة والاختزال في إزالة السموم من بعض الملوثات مثل البوليفينول ومواد الأنيلين من خلال البلمرة و/أو البلمرة المشتركة مع مركبات كيميائية اخرى أو من خلال التصاقها بأحماض الهيوميك (Zhou et al. 2022). تم إجراء عمليات إزالة السموم المتعددة لمركبات أو عناصر مختلفة مثل أصباغ الآزو والمعادن المشعة من خلال تفاعلات الأكسدة والاختزال (Al-Tohamy et al. 2020; Al-Tohamy et al. 2021; Ali et al.

Ali et al. 2021b; Ali et al. 2022a; Ye et al. 2021).

4.4. Oxygenases

The oxygenases enzymes are part of the oxidoreductase family that help for the oxidation of specific molecules through the transmission of O₂ from molecular oxygen. They are using NADH, NADPH or FAD as just co-substrates, and according to the quantity of oxygen atoms used for the oxygenation process; they are categorized as mono- or di-oxygenases (Gupta et al. 2021). They enhance the sensitivity and water solubility of the organic compounds and hence facilitating the aromatic ring breakdown. For instance, catechol ring was reported to be cleaved by the action of 1,2-dioxygenase enzyme produced from *Candida albicans* (Ali et al. 2022b). It has been mentioned that the interaction of the oxygen molecules with the organic pollutants through the oxygenase enzymes is the mean reason of the breakdown of their aromatic rings (Gangola et al. 2019).

2021b؛ Ali et al. 2022؛ Ye et al. 2021).

4.4. إنزيمات الأكسجين

تعد إنزيمات الأكسجين جزءًا من عائلة إنزيمات الأكسدة والاختزال التي تساعد على أكسدة جزيئات معينة من خلال نقل O_2 من الأكسجين الجزيئي. ويستخدمون NADH أو NADPH أو FAD كمجرد دعم مشترك، ووفقًا لكمية ذرات الأكسجين المستخدمة في عملية الأكسجة؛ يتم تصنيفها على أنها أحادي أو ثنائي الأكسجين (Gupta et al. 2021). وهي تعزز حساسية المركبات العضوية وقابليتها للذوبان في الماء وبالتالي تسهل تحلل الحلقات العطرية. على سبيل المثال، تم انشقاق حلقة الكاتيكول بفعل إنزيم 1,2-ديوكسيجيناز المنتج من *Candida albicans* (Ali et al. 2022b). وقد ذُكر أن تفاعل جزيئات الأكسجين مع الملوثات العضوية من خلال إنزيمات الأكسجين هو السبب لتحلل الحلقات العطرية (Gangola et al. 2019).

4.4.1. Monooxygenases

These enzymes are responsible for the implementation of single oxygen atom into the pollutant compounds, and according to the existence of their selective cofactor; they are classified into Flavin-dependent monooxygenases or P450 monooxygenases. Where NADP or NADPH or Flavin are the selective cofactors of Flavin-dependent monooxygenases. They are oxidative reaction enzymes for wide range of contaminants such as alkyl halides, fatty acids, and steroids (Hofrichter et al. 2020). According to their site selectivity and stereo-selectivity they have been classified as effective biocatalytic enzymes for the bioremediation of multipollutants, and have been recently used in synthetic chemistry (Abdel-Azeem et al. 2020).

Recent studies showed that a few monooxygenases can do their functions without cofactors, while the most of them should have cofactors. The monooxygenases that can work without cofactors are only need the molecular

4.4.1. إنزيمات الأكسجين الأحادية

هذه الإنزيمات مسؤولة عن إنشاء ذرة أكسجين واحدة في المركبات الملوثة، وبحسب وجود العامل المساعد الإنتقائي لها؛ يتم تصنيفها إلى أحادي الأكسجين، معتمد على الفلافين أو أحادي الأكسجين P450. حيث إن NADP أو NADPH أو Flavin هي العوامل المساعدة الإنتقائية لأحادي الأكسجين المعتمد على الفلافين. وهي إنزيمات تفاعل أكسدة لمجموعة واسعة من الملوثات مثل هاليدات الألكيل والأحماض الدهنية والستيرويدات (Hofrichter et al. 2020). ووفقًا لانتقائها في الموقع وانتقائها الاستريو، فقد تم تصنيفها على أنها إنزيمات تحفيزية حيوية فعالة للمعالجة الحيوية للملوثات المتعددة، وقد تم استخدامها مؤخرًا في الكيمياء الاصطناعية (Abdel-Azeem et al. 2020).

أظهرت الدراسات الحديثة أن عددًا قليلًا من إنزيمات الأكسجين الأحادية يمكن أن تؤدي وظائفها بدون عوامل مساعده، بينما يجب أن يكون لمعظمها عوامل مساعده. إن إنزيمات الأكسجين الأحادية التي يمكنها

oxygen to employ the pollutants as their oxidizing agents (Syed et al. 2021). The most interesting part of monooxygenases is their ability to catalyze some processes such as biodegradation, biotransformation, denitrification, dehalogenation, and ammonification of numerous aliphatic or aromatic compounds (Saxena et al. 2020). It has been reported that these enzymes can do oxidative dehalogenation reactions at O_2-rich conditions, while oxidizing dechlorination can only be occurred at O_2-low conditions (Hofrichter et al. 2020). Such advantages suggested the successful potential applications of monooxygenases in the biotransformation and biodegradation of aromatic pollutants (Abdel-Azeem et al. 2020).

4.4.2. Dioxygenases

They are enzymes that incorporate the molecular oxygen into their specific pollutants. They are mainly oxidizing the aromatic pollutants in order to clean them up from the ecosystem (Saxena et al. 2020). They are classified as non-heme Fe oxygenases family. In such family, everyone has one/two polypeptides as electron carriers before the oxygenase materials. In every alpha subunit, a cluster of 2Fe-2S structure in addition to mononuclear Fe has been detected (Xue et al. 2021). Catechol dioxygenase enzyme has been detected in soil microorganisms that playing an important role in the bioconversion of the aromatic precursors into aliphatic ones (Xue et al. 2021). It has been also detected that Fe (III) is used by the intradiol enzymes, while Mn (II) and Fe (II), in some cases, are used by extradiol enzymes (Narayanan et al. 2023).

4.5. Laccases

Laccases enzymes are classified as multi-copper oxidases that produced by many organisms including bacteria and fungi. They are catalyzing the oxidative degradation of multiple phenolic or aromatic pollutants with simultaneous reduction of molecular oxygen into water. They can be existed in different isoenzyme kinds depending on the distinct gene that is generally expressed according to the nature of the transcription factor in some cases (Behr et al. 2019). Laccases are secreted intracellularly or extracellularly by microbes can catalyze the oxidation of various compounds such as aryl diaminese, diphenols, polyphenols, polyamines, and inorganic salts (Mohapatra et al. 2022). They can target the methoxy groups of the methoxyphenolic acids or polyphenols, in addition to their ability to decracoxylate some functional groups, as well (Jha 2019; Khaliq 2023). Some factors can affect the specificity of the laccases towards their substrates such as pH changes, in addition to some chemicals that can block its action such as halides, hydroxides, cyanides, and azide structures (Valles et al. 2020), however, they proved potent biodegradation capacity to parathion, lignin, and glyphosate (Bhatt et al. 2023), in addition to lignocellulosic contaminants (Bhatt et al. 2022).

الماء. يمكن أن توجد في أنواع مختلفة من الإنزيمات المتجانسة، اعتمادًا على الجين المميز الذي يتم التعبير عنه عمومًا، وفقًا لطبيعة عامل النسخ في بعض الحالات (Behr et al. 2019). تُفرز اللاكيز داخل الخلايا أو خارجها بواسطة الميكروبات ويمكن أن يحفز أكسدة المركبات المختلفة مثل الأريل داي أمينيز، وثنائي الفينول، والبولي فينول، والبولي أمينات، والأملاح غير العضوية (Mohapatra et al. 2022). ويمكن أن تستهدف مجموعات الميثوكسي لأحماض الميثوكسي فينوليك أو البولي فينول، بالإضافة إلى قدرتها على نزع الكروكسيلات من بعض المجموعات الوظيفية، أيضًا (Jha 2019؛ خليج 2023). يمكن أن تؤثر بعض العوامل على خصوصية إنزيمات اللاكيز تجاه ركائزها مثل تغيرات الرقم الهيدروجيني، بالإضافة إلى بعض المواد الكيميائية التي يمكن أن تمنع عملها مثل الهاليدات والهيدروكسيدات والسيانيد وهياكل الأزيد (Valles et al. 2020)، ومع ذلك، فقد أثبتوا قدرة تحلل بيولوجي قوية للباراثيون واللجنين والجليفوسات (Bhatt et al. 2023)، بالإضافة إلى الملوثات اللجنوسليلوزية (Bhatt et al. 2022).

4.6. Cytochrome P450

They are heme enzymes that present in all biological entities including Eukaryotes and Prokaryotes (Li et al. 2020). They are playing an essential role in the bioremediation mechanisms as they have potential destroying capacity via the biotransformation of xenobiotics such as dehalogenation, aliphatic hydroxylations, and dealkylations, in addition to their ability to inactivate numerous metabolic pathways (Narayanan et al. 2023). Some molecules such as ferredoxin reductase and ferredoxin are almost acting as electron sources for the activity of the enzymes. They have been applied for the bioremediation of organic pollutants and hydrocarbon products depending of using microbially engineered forms of their proteins (Bhandari et al. 2021). Generally, some of the microbes and their enzymes used for the bioremediation of various environmental pollutants are summarized in Table 2.

4.6. سيتوكروم P450

وهي إنزيمات الهيم الموجودة في جميع الكيانات البيولوجية بما في ذلك حقيقيات وبدائيات النواة (Li et al. 2020). وهذه الإنزيمات تلعب دورًا أساسيًا في آليات المعالجة الحيوية، حيث إن لديهم قدرة تدمير محتملة من خلال التحول الحيوي للزينوبيوتيك مثل إزالة الهالوجين، والهيدروكسيل الأليفاتي، بالإضافة إلى قدرتهم على تعطيل العديد من مسارات الأيض (Narayanan et al. 2023). تعمل بعض الجزيئات مثل اختزال فيريدوكسين وفيريدوكسين كمصادر إلكترونية لنشاط الإنزيمات. وقد تم تطبيقها على المعالجة الحيوية للملوثات العضوية والمنتجات الهيدروكربونية اعتمادًا على استخدام أشكال معدلة ميكروبيا من بروتيناتها (Bhandari et al. 2021). بشكل عام، تم تلخيص بعض الميكروبات وإنزيماتها المستخدمة للمعالجة الحيوية لمختلف الملوثات البيئية في الجدول 2.

Table 2 Microbes and their enzymes used for the bioremediation of various pollutants in the environment (Ayilara and Babalola 2023).

الجدول 2 الميكروبات وإنزيماتها المستخدمة في المعالجة الحيوية لمختلف الملوثات في البيئة (Ayilara and Babalola 2023).

Microbial source	Enzyme type	Type of pollutant	References
Bacillus pumilus	Lipase	Palm oil زيت النخيل	(Saranya et al. 2019)
E. coli	Dehydrogenase	Steroids منشطات	(Ye et al. 2019)
Bacillus pumilus	Lipase	Oil-contaminated industrial wastewater مياه الصرف الصناعي الملوثة بالنفط	(Saranya et al. 2019)
Brevundimonas diminuta	Phosphotriesterase	Pesticides المبيدات الحشرية	(Thakur et al. 2019)
Pseudomonas sp.	Oxygenase	Pesticides المبيدات الحشرية	(Malakar et al. 2020)
Bacillus safenis	Oxidoreductase	Xenobiotics الزينوبيوتيك	(Malakar et al. 2020)
Bacillus subtilis	Protease	Casein and feather الكازين والريش	(Bhandari et al. 2021)
Escherichia coli and *Bacillus* sp. F31	Lignin peroxidase	Synthetic-dyes الأصباغ المصنعة	(Dave and Das 2021)
Pseudomonas putida	Laccase	Synthetic-dyes الأصباغ المصنعة	(Bhandari et al. 2021)
T. fusca	Hydrolases	Polyester plastics بلاستيك البوليستر	(Gricajeva et al. 2022)

5. Genetic modification and engineering for enhanced bioremediation

Genetic engineering technologies have been performed to boost the ability of using enzymes in an industrial scale in the environmental applications, since the using of native microbial enzymes are economically un-applicable. Various genetic experiments such as in vitro gene mutagenesis and expression can achieve such target (Saravanan et al. 2021).

Some genetic engineering-advanced methods have been developed in order to boost the control of environmental pollution such as the following:

1- Investigation and cloning of the promising effective genes that assist for the improvement of the bioremediation process.
2- Boosting the expression of the pollutants degrading enzymes.
3- Construction of super-engineered microbes having the genes

5. التعديل الوراثي والهندسة الوراثية لتعزيز المعالجة الحيوية

تم إجراء تقنيات الهندسة الوراثية لتعزيز القدرة على استخدام الإنزيمات على نطاق صناعي في التطبيقات البيئية، حيث إن استخدام الإنزيمات الميكروبية الأصلية غير قابل للتطبيق اقتصاديًا. يمكن أن تحقق التجارب الجينية المختلفة مثل الطفرات والتعبيرات الجينية في المختبر هذا الهدف (Saravanan et al. 2021).

تم تطوير بعض الطرق المتقدمة في مجال الهندسة الوراثية من أجل تعزيز مكافحة التلوث البيئي مثل ما يلي:

1- العمل على استنساخ الجينات الفعالة الواعدة التي تساعد على تحسين عملية المعالجة الحيوية.
2- تعزيز التعبير عن الإنزيمات المحللة للملوثات.
3- العمل على وجود ميكروبات مهندسة وراثيا تحمل الجينات المسؤولة عن تحلل الملوثات المختلفة.
4- اندماج البروتوبلاست بين الخلايا الأصلية والخلايا

responsible for various pollutants degradation.
4- Protoplast fusion between the native and the engineered cells in order to combine the advantages of both cells (Figure 6).

The development of efficient degrading genetically-modified bacteria has been achieved through the identification of the degradation pathways, genes, and the degradation mechanisms of the microbial cells. Huy and his team succeeded to extract the cDNA of the laccase gene from *Fusarium oxysporum* with subsequent cloning and expression in *Pichia pastoris*, as it showed potent bioremediation activity against remazol blue, methyl orange, aniline blue, and indigo carmine dyes (Huy et al. 2021). Similarly, multiple genes including laccase, MnP, and LiP enzymes were cloned from *Aspergillus* sp. into *Pichia pastoris* to be used for effective bioremediation of Congo red (Liu et al. 2020).

المهندسة وراثيا من أجل الجمع بين مزايا كلتا الخليتين (الشكل 6).

وقد تم تطوير بكتيريا فعالة ومعدلة وراثيًا من خلال تحديد مسارات التحلل، والجينات، وآليات التحلل للخلايا الميكروبية. نجح Huy وفريقه في استخراج الحمض النووي cDNA لجين اللاكيز من *Fusarium oxysporum* مع الاستنساخ والتعبير اللاحقين في *Pichia Pastoris* ، حيث أظهر نشاطًا قويًا في المعالجة الحيوية ضد أصباغ الريمازول الأزرق، والبرتقالي الميثيلي، والأزرق الأنيليني، والكارمين النيلي (.Huy et al 2021). وبالمثل، تم استنساخ جينات متعددة بما في ذلك إنزيمات اللاكيز، MnP، و LiP من *Aspergillus* sp إلى *Pichia Pastoris* لاستخدامها في المعالجة الحيوية الفعالة لأحمر الكونغو (Liu et al. 2020).

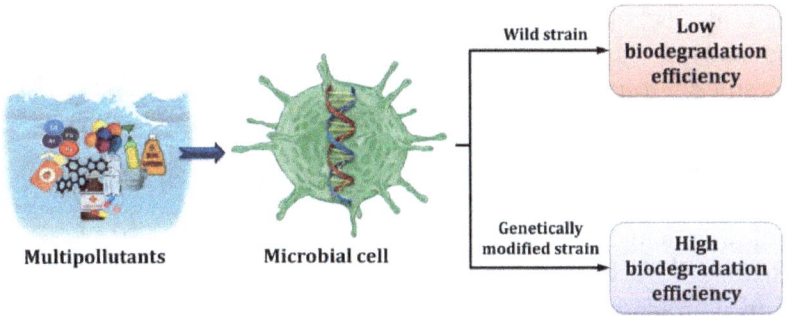

Figure 6 Effective bioremediation of multipollutants using genetically engineered microbial cells (Narayanan et al. 2023).

الشكل 6 المعالجة الحيوية الفعالة للملوثات المتعددة باستخدام الخلايا الميكروبية المعدلة وراثيًا (Narayanan et al. 2023).

The gene cloning and editing are important traits of the genetic engineering field that has been used for the enhancement of bioremediation of different waste in the environment (Figure 7). It has activated the microorganisms to transform specific pollutants from toxic forms into non-toxic forms, especially those contaminants which are hardly biodegraded (Kaur and Sodhi 2022). Due to genetic engineering, scientists now, can increase the potency and effectiveness of the capacity of microorganisms to breakdown the environmental contaminants. PET is a common plastic material used in the preparation of different medical equipment pieces, is

يُعد استنساخ الجينات وتحريرها من السمات المهمة في مجال الهندسة الوراثية الذي تم استخدامه لتعزيز المعالجة الحيوية للنفايات المختلفة في البيئة (الشكل 7). لقد قام Kaur and Sodhi بتنشيط الكائنات الحية الدقيقة لتحويل ملوثات محددة من الأشكال السامة إلى أشكال غير سامة، خاصة تلك الملوثات التي نادرًا ما تتحلل بيولوجيًا (Kaur and Sodhi 2022). بفضل الهندسة الوراثية، أصبح بإمكان العلماء الآن زيادة قوة وفعالية قدرة الكائنات الحية الدقيقة على تكسير الملوثات البيئية. البولي إيثيلين تيرفثالات (PET) هو مادة بلاستيكية شائعة تُستخدم في تحضير قطع المعدات الطبية المختلفة،

an example of how genetic engineering help for its bioremediation. It is a biodegradation resistant substance that can stay in the environment for long times without changes. In a recent research, PET was discovered to be degraded by a bacterial species known as *Ideonella sakaiensis*. The bacterial capacity to degrade PET was improved through modifying its DNA using genetic engineering tools (Ebah et al. 2022). In a similar study, the fungus *Phanerochaete chrysosporium* was genetically modified in order to increase its capability to degrade polycyclic aromatic hydrocarbons, which are considered hazardous substances that can develop cancers in humans (Alao and Adebayo 2022; Srivastava and Kumar 2019).

وهو مثال على كيفية مساعدة الهندسة الوراثية في معالجتها بيولوجيًا. وهي مادة مقاومة للتحلل الحيوي ويمكن أن تبقى في البيئة لفترات طويلة دون تغيير. في بحث أجري مؤخرًا، تم اكتشاف أن مادة PET تتحلل بواسطة أنواع بكتيرية تعرف باسم *Ideonella sakaiensis*. تم تحسين القدرة البكتيرية على تحليل PET من خلال تعديل الحمض النووي الخاص بها باستخدام أدوات الهندسة الوراثية (Ebah et al. 2022). في دراسة مماثلة، تم تعديل فطر *Phanerochaete chrysosporium* وراثيا من أجل زيادة قدرته على تحليل الهيدروكربونات العطرية متعددة الحلقات، والتي تعتبر مواد خطرة يمكن أن تسبب السرطان لدى البشر (Alao and Adebayo 2022; Srivastava and Kumar 2019).

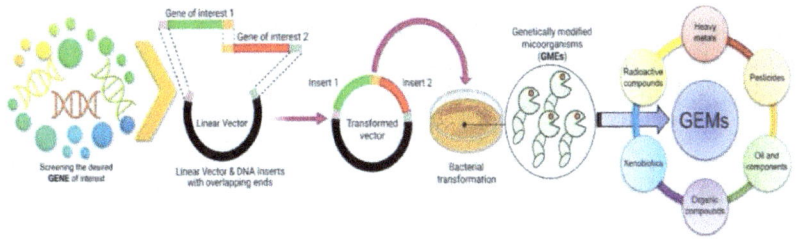

Figure 7 Using of genetic engineering as a tool for the enhancement of the bioremediation of different environmental pollutants (Chandra et al. 2023).

الشكل 7 استخدام الهندسة الوراثية كأداة لتعزيز المعالجة الحيوية للملوثات البيئية المختلفة (Chandra et al. 2023).

It has been also reported that the presence of transposons, plasmids, and chromosomal genes that are encoding the catabolic enzymes such as transferases, dioxygenases, hydrolases, isomerases, etc., are providing the power to the microbes in order to effectively degrade xenobiotics (Bhatt et al. 2021a). The genetic exchange and genetic rearrangements are what the genes' localization is based on. On the other hand, most catabolic genes are usually found on the plasmids and other genetically elements that can be mobile, and hence, the ability of an organism to degrade a specific pollutant is depending on the absence or presence of the plasmid that is carrying the genes responsible for

تم عمل دراسة عن وجود الترانسبوزونات والبلازميدات وجينات الكروموسومات التي تشفر الإنزيمات الكاتابولية مثل الترانسفيراز، وثنائي الأكسجين، والهيدرولازات، والإيزومراز، وما إلى ذلك، توفر الطاقة للميكروبات من أجل تحلل الزينوبيوتيك بفعالية (Bhatt et al. 2021a). التبادل وإعادة الترتيب الجيني هما ما يعتمد عليه توطين الجينات. من ناحية أخرى، فإن معظم الجينات الكاتابوليك عادةً ما توجد على البلازميدات والعناصر الوراثية الأخرى التي يمكن أن تكون متحركة، وبالتالي، فإن قدرة الكائن الحي على تحليل ملوث معين تعتمد على غياب أو وجود البلازميد الذي يحمل الجينات المسؤولة عن تحليل هذا الملوثات

degrading of such pollutant inside the microbial cells (Narayanan et al. 2023).

6. Types of Environmental Wastes
6.1. Wastewater

The removal of pollutants from water, which is the most essential entity for life, is crucial. The recent industrial renaissance has put a huge pressure on the using of water as it is an essential compound in the production process. The polluted industrial effluents are increasing when the production of industrial substances increase. Cost-effective strictly ways are required for the treatment of such industrial effluents in order to keep the sustainable industrial development and to keep the environment in a safe condition. The reduction of the toxicity of these effluents has been applied through various electrochemical, valorization, and advanced oxidation techniques (Gupta and Shukla 2020).

But due to the invalidity of these techniques for all industries; scientists incorporated the nanoscience and nanotechnology fields in

داخل الخلايا الميكروبية (Narayanan et al. 2023).

6. أنواع النفايات البيئية
6.1. مياه الصرف

إن إزالة الملوثات من الماء، وهو الكيان الأكثر أهمية للحياة، قد فرضت من خلال النهضة الصناعية الحديثة ضغطًا كبيرًا على استخدام المياه لأنها ماده أساسية في عملية الإنتاج. تزداد النفايات الصناعية الملوثة عندما يزداد إنتاج المواد الصناعية. ويلزم وجود طرق صارمة فعالة من حيث التكلفة لمعالجة هذه النفايات الصناعية السائلة من أجل الحفاظ على التنمية الصناعية المستدامة والحفاظ على البيئة في حالة آمنة. وقد تم تطبيق الحد من سمية هذه النفايات السائلة من خلال العديد من التقنيات الكهروكيميائية والتثمينية وتقنيات الأكسدة المتقدمة (Gupta and Shukla 2020).

ولكن نظرا لعدم صلاحية هذه التقنيات لجميع الصناعات؛ قام العلماء بدمج مجالات علوم النانو وتكنولوجيا النانو في تنظيف مياه الصرف الصحي كتقنيات

wastewater cleaning as recent, cost-effective and developed technologies. The produced nanomaterials have superior chemical properties, small size, and high surface area to volume ratio properties compared with their conventional counterparts (Sakshi and Bharadvaja 2023). The ability of microorganisms and their extracts to synthesis nanomaterials is a green method that paved the way toward the eco-friendly and cost-effective remediation of pollutants (Bolade et al. 2020).

In some cases, the enhancement of the ability of these nanomaterials to remediate pollutants is accomplished by their integration into polymeric membranes. As they improve the membrane permeability, its temperature and mechanical strength, in addition to pollutant degradation through their present innovative functions. Moreover, the metal-organic frameworks (MOFs) have also showed potent ability to exclude heavy metals from wastewater, as they structure is synthesized through the coordination between organic ligands and metal ion

حديثة وفعالة من حيث التكلفة والتطور. تتمتع المواد النانوية المنتجة بخصائص كيميائية فائقة، وحجم صغير، ومساحة سطحية عالية لخصائص نسبة الحجم مقارنة بنظيراتها التقليدية (Sakshi and Bharadvaja 2023). إن قدرة الكائنات الحية الدقيقة ومستخلصاتها على تخليق المواد النانوية هي طريقة خضراء مهدت الطريق نحو معالجة الملوثات الصديقة للبيئة والمنخفضة التكلفة (Bolade et al. 2020).

في بعض الحالات، يتم تعزيز قدرة هذه المواد النانوية على معالجة الملوثات من خلال دمجها في الأغشية البوليمرية. وبينما تعمل على تحسين نفاذية الغشاء ودرجة حرارته وقوته الميكانيكية، بالإضافة إلى تحلل الملوثات من خلال وظائفها المبتكرة الحالية. علاوة على ذلك، أظهرت أطر العمل المعدنية العضوية (MOFs) أيضًا قدرة قوية على استبعاد المعادن الثقيلة من مياه الصرف الصحي، حيث يتم تخليقها من خلال التنسيق بين الروابط العضوية وسلائف الأيونات المعدنية (Deshpande et al. 2020).

precursors (Deshpande et al. 2020).

However, from a biological point of view, there is now a brand name product known as Effective microorganisms (EM) that has been developed by the Japanese Scientist Dr. Teruo Higa. The product is a liquid one that contains aerobic and anaerobic nonpathogenic microorganisms. It is an eco-friendly solution that naturally fermented without chemical or genetic engineering intervention. It contains five families and ten genera of beneficial microorganisms beside other dozens of types of lactic acid bacteria, actinomycetes, yeast, and photosynthetic bacteria that can all existed together at pH level of 3.5. This EM mixture was basically prepared to be used in organic farming, however, it has gain maximum interest in recent years to be used in wider applications including medicine, agriculture, livestock, and environmental sectors. It terms of environmental protection, it can be used in waste deodorization, water quality improving, sludge treatment, composting, and wastewater

ومع ذلك، من وجهة نظر بيولوجية، يوجد الآن منتج يحمل اسمًا تجاريًا يُعرف باسم الكائنات الحية الدقيقة الفعالة (EM) تم تطويره بواسطة العالم الياباني .Dr Teruo Higa. المنتج عبارة عن سائل يحتوي على كائنات حية دقيقة لاهوائية ولاهوائية غير مسببة للأمراض. وهو محلول صديق للبيئة يتم تخميره بشكل طبيعي دون تدخل هندسة كيميائية أو وراثية. وهو يحتوي على خمس عائلات وعشرة أجناس من الكائنات الحية الدقيقة المفيدة إلى جانب عشرات الأنواع الأخرى من بكتيريا حمض اللاكتيك والكتينوميسيت والخميرة والبكتيريا الضوئية التي يمكن أن تتواجد جميعها معًا عند مستوى حموضة يبلغ 3.5. وقد تم إعداد خليط EM في الأسواق الناشئة بشكل أساسي لاستخدامه في الزراعة العضوية، ومع ذلك، فقد اكتسب أقصى قدر من الاهتمام في السنوات الأخيرة لاستخدامه في التطبيقات الأوسع نطاقًا بما في ذلك الطب والزراعة والثروة الحيوانية والقطاعات البيئية. من حيث حماية البيئة، يمكن استخدامه في إزالة الروائح الكريهة من النفايات وتحسين جودة

treatment (Safwat and Matta 2021).

Microalgae and cyanobacteria have been also used for the bioremediation of water contains pesticides and heavy metals. In a recent study, the authors used combined microalgae and cyanobacteria strains such as *Spirulina platensis*, *Scenedesmus quadricuda*, and *Chlorella vulgaris*, to bioremediate malathion pesticide and the heavy metals lead, cadmium, and nickel from the urban wastewater and agricultural drainage water. At the end of the experiment, it has been reported the removal of 99% of malathion and 95% of nickel effectively by microalgae with lower bit activity towards the bioaccumulation of lead and cadmium (Abdel-Razek et al. 2019).

المياه ومعالجة الحمأة والتحويل إلى سماد ومعالجة مياه الصرف الصحي (Safwat and Matta 2021).

كما تم استخدام الطحالب الدقيقة والبكتيريا الزرقاء للمعالجة البيولوجية للمياه التي تحتوي على مبيدات حشرية ومعادن ثقيلة. في دراسة حديثة، استخدم المؤلفون سلالات الطحالب الدقيقة والبكتيريا الزرقاء المجمعة مثل *Spirulina Platensis*، و*Scendesmus Quadricuda*، و*Chlorella vulgaris* للعلاج الحيوي لمبيدات الملاثيون الحشرية والرصاص والمعادن الثقيلة والكادميوم والنيكل من مياه الصرف الصحي في المناطق الحضرية ومياه الصرف الزراعي. في نهاية التجربة، تم التوصل الى نتائج تشير الى إزالة 99% من الملاثيون و95% من النيكل بشكل فعال بواسطة الطحالب الدقيقة مع نشاط أقل تجاه التراكم الحيوي للرصاص والكادميوم (Abdel-Razek et al. 2019).

6.1.1. Factors cause water pollution
6.1.1.1. Inorganic compounds

Many inorganic compounds may be released to the water bodies through different human activities such as unsafe agricultural practices, industrial discharges, and inadequate sanitation systems. These inorganic chemicals include heavy metals, inorganic anions, metal-halides, radioactive-elements, trace elements, cations, inorganic salts, oxyanions, cyanides, etc., (Srivastav and Ranjan 2020). Most of these inorganic pollutants are not biodegradable or thermo-degradable and hence remain in the environment for long times, which almost causes further environmental harmfulness (Wasewar et al. 2020; Kumar et al. 2020b). In addition, most toxic heavy metals are released to the environment through the industrial wastewater, which harm the surrounding ecosystem and its living organisms. Multiple activities such as hospital waste, electroplating, mines, battery-plants, electronic factories, alloys, and smelters are big sources of toxic heavy metals.

6.1.1 العوامل المسببة لتلوث المياه
6.1.1.1. المركبات غير العضوية

قد تتسرب العديد من المركبات غير العضوية إلى المسطحات المائية من خلال أنشطة بشرية مختلفة مثل الممارسات الزراعية غير الآمنة والتصريفات الصناعية وأنظمة الصرف الصحي غير الكافية. تشمل هذه المواد الكيميائية غير العضوية المعادن الثقيلة، والأيونات غير العضوية، وهاليدات المعادن، والعناصر المشعة، والعناصر النادرة، والكاتيونات، والأملاح غير العضوية، والأوكسيونات، والسيانيد، وما إلى ذلك، (Srivastav and Ranjan 2020). ومعظم هذه الملوثات غير العضوية غير قابلة للتحلل الحيوي أو التحلل الحراري ومن ثم تبقى في البيئة لفترات طويلة، مما يتسبب تقريبًا في مزيد من الضرر البيئي (Wasewar et al. 2020; Kumar et al. 2020b). بالإضافة إلى ذلك، تنطلق معظم المعادن الثقيلة السامة إلى البيئة من خلال مياه الصرف الصناعي، والتي تضر بالنظام البيئي المحيط والكائنات الحية. تُعد الأنشطة المتعددة مثل مخلفات المستشفيات والطلاء الكهربائي والمناجم ومصانع

However, in some cases these heavy metals might be polluted by other natural or anthropogenic causes such as volcanoes, rock disintegration, and soil erosion (Srivastav and Ranjan 2020; Ojha et al. 2021).

6.1.1.2. Organic compounds

Organic contaminants are almost released to water through the industrial and/or agricultural activities. These activities may cause the release of wastewater loaded with high concentrations of pesticides or herbicides or fertilizers from farmlands, or high concentrations of polyaromatic hydrocarbons from coke plant, or various heterogeneity compounds such as PCB, PBDE from chemical industries, or other organic pollutants released from food industry or from municipal.

All these pollutants are extreme potential cause of human health hazards in addition to their bad effect on the environment (Wasewar et al. 2020).

البطاريات والمصانع الإلكترونية والسبائك والمصاهر مصادر كبيرة للمعادن الثقيلة السامة. ومع ذلك، في بعض الحالات، قد تتلوث هذه المعادن الثقيلة لأسباب طبيعية أو بشرية أخرى مثل البراكين وتفكك الصخور وتآكل التربة (Srivastav and Ranjan 2020; Ojha et al. 2021).

6.1.1.2. المركبات العضوية

تتسرب الملوثات العضوية تقريبًا إلى المياه من خلال الأنشطة الصناعية و/أو الزراعية. قد تتسبب هذه الأنشطة في إطلاق مياه الصرف الصحي المحملة بتركيزات عالية من المبيدات الحشرية أو مبيدات الأعشاب أو الأسمدة من الأراضي الزراعية، أو تركيزات عالية من الهيدروكربونات متعددة الأرومات من مصانع فحم الكوك، أو مركبات غير متجانسة مختلفة مثل ثنائي الفينيل متعدد الكلور، والإثير متعدد البروم ثنائي الفينيل من الصناعات الكيميائية، أو ملوثات عضوية أخرى يتم تسريبها من صناعة الأغذية أو من البلدية. وتعتبر جميع هذه الملوثات هي سبب محتمل للمخاطر الصحية على الإنسان بالإضافة إلى تأثيرها السيء على البيئة (Wasewar et al. 2020).

6.1.2. Microbial-assisted nanomaterials for wastewater treatment

Recent studies showed the ability of different microbes such as bacteria and fungi in addition to plants to produce sustainable and eco-friendly nanomaterials. These green synthesized nanoparticles can be potential alternatives for the chemically synthesized nanomaterials that may have some disadvantages (Mandeep and Shukla 2020). However, both chemically and biologically synthesized nanoparticles share a lot of advantages. They are small in size, have specific physical/chemical/biological properties that make them suitable for use in various applications including wastewater treatment. They can represent photocatalytic degradation or adsorption capabilities in addition to other various nano-techniques that help them to remediate the industrial effluents effectively (Palit and Hussain 2020).

The green synthesized nanoparticles have an extra advantage of being able to add to their outer surfaces some of the proteinaceous and

6.1.2. المواد النانوية بمساعدة الميكروبات لمعالجة مياه الصرف

أظهرت الدراسات الحديثة قدرة الميكروبات المختلفة مثل البكتيريا والفطريات بالإضافة إلى النباتات على إنتاج مواد نانوية مستدامة وصديقة للبيئة. يمكن أن تكون هذه الجسيمات النانوية الخضراء المُخلقة بدائل محتملة للمواد النانوية المخلقه كيميائيًا التي قد يكون لها بعض العيوب (Mandeep and Shukla 2020). ومع ذلك، تشترك الجسيمات النانوية المخلقة كيميائيًا وبيولوجيًا في الكثير من المزايا. وهي صغيرة الحجم ولها خصائص فيزيائية/كيميائية/بيولوجية محددة تجعلها مناسبة للاستخدام في مختلف التطبيقات بما في ذلك معالجة مياه الصرف. ويمكن أن تمثل قدرات التحلل الضوئي أو الامتزاز بالإضافة إلى تقنيات النانو المختلفة التي تساعدهم على معالجة النفايات السائلة الصناعية بشكل فعال (Palit and Hussain 2020).

تتمتع الجسيمات النانوية الخضراء بميزة إضافية تتمثل في قدرتها على إضافة بعض العناصر البروتينية والعناصر النشطة

bioactive elements from the microbial media they formed in. Iron oxide nanoparticles have been synthesized by *Aspergillus tubingensis* (STSP 25), and were able to remove many heavy metals such as Pb^{2+}, Ni^{2+}, Cu^{2+}, and Zn^{2+} from the wastewater with a total percentage of more than 90% for five successive cycles (Mahanty et al. 2020). Moreover, some nanoparticles can co-precipitate with biopolymers produced in the microbial extract and used for different applications. For instance, Govarthanan and his team used the exopolysaccharides produced by *Chlorella vulgaris* to co-precipitate iron oxide nanoparticles in order to form a nanocomposite that was able to remove ammonium and phosphate ions with percentages 85 and 91%, respectively (Govarthanan et al. 2020). *Escherichia* sp. SINT7 has also been used for the biosynthesis of copper nanoparticles that were able to degrade azo dyes and textile effluents. They were able to reduce the concentrations of direct blue-1, Congo red, black-5, and malachite green at its lower concentration (25 mg/L). They were also able to

بيولوجيًا إلى أسطحها الخارجية من الوسائط الميكروبية التي تكونت فيها. تم تصنيع الجسيمات النانوية لأكسيد الحديد بواسطة *Aspergillus Pipeensis* (STSP 25)، وكانت قادرة على إزالة العديد من المعادن الثقيلة مثل Pb^{2+}, Ni^{2+}, Cu^{2+}, Zn^{2+} من مياه الصرف بنسبة إجمالية تزيد عن 90% لخمس دورات متتالية (Mahanty et al. 2020). علاوة على ذلك، يمكن لبعض الجسيمات النانوية أن تترسب مع البوليمرات الحيوية المنتجة في المستخلص الميكروبي وتستخدم في تطبيقات مختلفة. على سبيل المثال، استخدم Govarthanan وفريقه مركبات عديدة التسكر الخارجية التي تنتجها *Chlorella vulgaris* للتكسير المشترك لجسيمات أكسيد الحديد النانوية من أجل تكوين مركب نانوي قادر على إزالة أيونات الأمونيوم والفوسفات بنسبة 85 و91%، على التوالي (Govarthanan et al. 2020). كما تم استخدام *Escherichia* sp. SINT7 للتخليق الحيوي لجسيمات النحاس النانوية التي كانت قادرة على تحلل أصباغ الآزو مخلفات النسيج السائلة. وكانت قادرة أيضا على تقليل تركيزات الأزرق-1 المباشر،

reduce the suspended solids and the phosphate and chloride ions in the treated wastewater samples (Noman et al. 2020).

In another study, *Pseudoalteromonas* sp. CF10-13 has been used for the preparation of iron-sulfur nanoparticles in order to be used for the degradation of Napthol Green B dye depending on the extracellular transfer of the electrons (Cheng et al. 2019). Moreover, the supernatant lacks *Pseudomonas aeruginosa* has been used for the production of Zirconia nanoparticles that has been effectively used for the adsorption of 526.32 mg/g of Tetracycline (Debnath et al. 2020).

Another indirect way can be used in order to maximize the benefits of using microorganisms in environmental biotechnology that depends on nanotechnology. For instance, microorganisms can help for the bioremediation of effluents through the production of catalytic enzymes that can be bind to nanoparticles and can effectively remediate the environmental pollutants.

والأحمر الكونغوي، والأسود-5، والأخضر المالاكيت بتركيزات أقل (25 مجم/لتر). كما تمكنت من تقليل المواد الصلبة العالقة وأيونات الفوسفات والكلوريد في عينات مياه الصرف المعالجة (Noman et al. 2020).

في دراسة أخرى، تم استخدام CF10-13 *Pseudoalteromonas* sp لتحضير جسيمات الحديد والكبريت النانوية من أجل استخدامها لتحليل صبغة Napthol Green B اعتمادًا على الإلكترونات خارج الخلية (Cheng et al. 2019). علاوة على ذلك، فان *Pseudomonas aeruginosa* تم استخدامها لإنتاج جسيمات Zirconia النانوية والتي كان لها دور فعال لامتصاص 526.32 مجم/جم من التتراسيكلين (Debnath et al. 2020).

ويمكن استخدام طريقة أخرى غير مباشرة لتعظيم فوائد استخدام الكائنات الحية الدقيقة في التكنولوجيا الحيوية البيئية التي تعتمد على تكنولوجيا النانو. على سبيل المثال، يمكن أن تساعد الكائنات الحية الدقيقة في المعالجة الحيوية للمخلفات السائلة من خلال إنتاج الإنزيمات التحفيزية التي يمكن

6.1.3. Enzyme-integrated nanoparticles for wastewater treatment

The integration between enzymes and nanoparticles has mutual advantages for both technologies. For nanomaterials, it could be converted into less harmful particles to the surrounding environment. The presence of enzymes combined with nanoparticles is also minimizing the cell interaction through steric hindrances, in addition to decreasing their surface energy, and makes nanoparticles more adaptable and more efficient in bioremediation process (Mandeep and Shukla 2020). On the other hand, for enzymes, they become highly stable due to the prevention of their unfolding, and they also become less vulnerable to diffusional-constraints, and can be used for several successive cycles, besides being have improved kinetic characteristics. The immobilization of enzymes into the surface of nanoparticles increases the amount of the enzymes loaded to the surface due to the large

6.1.3. جسيمات نانوية مدمجة بالإنزيمات لمعالجة مياه الصرف

إن التكامل بين الإنزيمات والجسيمات النانوية له مزايا متبادلة لكلا التقنيتين. بالنسبة للمواد النانوية، يمكن تحويلها إلى جسيمات أقل ضررًا على البيئة المحيطة. كما أن وجود الإنزيمات جنبًا إلى جنب مع الجسيمات النانوية يقلل أيضًا من تفاعل الخلايا من خلال العوائق الفراغية، بالإضافة إلى تقليل الطاقة السطحية، ويجعل الجسيمات النانوية أكثر قابلية للتكيف وأكثر كفاءة في عملية المعالجة الحيوية (Mandeep and Shukla 2020). من ناحية أخرى، بالنسبة للإنزيمات، تصبح مستقرة للغاية بسبب منع تكشفها، كما تصبح أقل عرضة لقيود الانتشار، ويمكن استخدامها لعدة دورات متتالية، بالإضافة إلى أن لها خصائص حركية محسنة. يؤدي تثبيت الإنزيمات في سطح الجسيمات النانوية إلى زيادة كمية الإنزيمات المحملة على السطح بسبب المساحة السطحية الكبيرة للمواد النانوية، كما يساعد على فصل الإنزيمات بسهولة

ربطها بالجسيمات النانوية ويمكنها معالجة الملوثات البيئية بشكل فعال.

surface area of the nanomaterials, and also help for the easily separation of the enzymes from the reaction mixture (Mandeep and Shukla 2020).

In the same context, Darwesh and his colleagues succeeded to immobilize the peroxidase enzyme on iron oxide nanoparticles modified with glutaraldehyde in order to be used in wastewater bioremediation. They found that the nanoparticles increased the pH and/or the temperature stability of the immobilized enzyme. Their study showed the potent ability of the peroxidase enzyme to individually remediate the green and red azo dyes in 4 h. However, it also succeeded to remove both combined dyes in 6 h, completely (Darwesh et al. 2019).

In another study, laccase immobilized Fe_3O_4-chitosan composite was used to effectively degrade 2,4-Dicholor-Phenol and 4-Choloro-Phenol for up to 10 cycles with percentages of 75.5 and 91.4%, respectively (Zhang et al. 2020b). Moreover, laccase enzyme has

عن خليط التفاعل (Mandeep and Shukla 2020).

وفي السياق نفسه، نجح Darwesh وزملاؤه في تثبيت إنزيم بيروكسيداز على جسيمات أكسيد الحديد النانوية المعدلة بجلوتارالدهيد لاستخدامها في المعالجة الحيوية لمياه الصرف. ووجدوا أن الجسيمات النانوية زادت من درجة الحموضة و/أو ثبات درجة حرارة الإنزيم المثبت. وأظهرت دراستهم القدرة القوية لإنزيم بيروكسيداز على معالجة صبغات الآزو الخضراء والحمراء بشكل فردي في 4 ساعات. ومع ذلك، نجح أيضًا في إزالة كلتا الصبغتين المجمّعتين في 6 ساعات، تمامًا (Darwesh et al. 2019).

في دراسة أخرى، تم استخدام مركب Fe_3O_4-chitosan المثبّت في انزيم اللاكيز وذلك لتحليل 2,4--Dicholor Phenol و4-Choloro-Phenol بفعالية لمدة تصل إلى 10 دورات بنسبة 75.5 و91.4%، على التوالي (Zhang et al. 2020b). علاوة على ذلك، تم أيضًا تثبيت

been also immobilized to iron oxide core and chelated Cu^{2+} of carbon shell. The immobilized enzyme showed potent remediation activity against reactive blue 19axophloxine, , procion red MX-5B, crystal violet, malachite green, and brilliant green dyes, with reusability up to 10 continuous cycles (Li et al. 2020). Similarly, Guo and his team succeeded to immobilize lignin peroxidase enzyme on the surface of Fe_3O_4@SiO_2@polydopamine nanocomposite. Comparing with the free enzyme, the immobilized enzyme showed degradation capability of 100% toward phenol, dibutyl phthalate, tetracycline, and 5-chlorophenol (Guo et al. 2019).

6.2. Solid waste

Solid waste has become a global serious problem since the recent rapid urbanization and the economic development. These wastes are originating from various sources such as industrial, municipal, agricultural, and domestic sources (Zhou et al. 2019). The domestic solid wastes are those which are collected from streets, country parks, litter bins, and residential buildings. It has

إنزيم اللاكيز في لب أكسيد الحديد ومادة Cu_2. أظهر الإنزيم المثبت نشاطًا علاجيًا قويًا ضد 19 الأصباغ الزرقاء التفاعلية 19أكسوفلوكسين، والبروسيون الأحمر MX-5B ، البنفسجي البلوري ، والأخضر الملكيت، والأصباغ الخضراء اللامعة، مع إمكانية إعادة الاستخدام حتى 10 دورات مستمرة (Li et al. 2020). وبالمثل، نجح Guo وفريقه في تثبيت إنزيم الليجنين بيروكسيداز على سطح مركب Fe_3O_4@SiO_2@بوليدوبامين نانوي. وبالمقارنة مع الإنزيم الغير مثبت، أظهر الإنزيم المثبت قدرة تحليل بنسبة 100% تجاه الفينول، وثنائي بيوتيل الفثالات، وتتراسيكلين، و5-كلوروفينول (Guo et al. 2019).

6.2. المخلفات الصلبة

أصبحت المخلفات الصلبة مشكلة خطيرة عالمية منذ التحضر السريع والتنمية الاقتصادية مؤخرًا. تنشأ هذه المخلفات من مصادر مختلفة مثل المصادر الصناعية والبلدية والزراعية والمحلية (Zhou et al. 2019). المخلفات الصلبة المنزلية هي تلك التي يتم جمعها من الشوارع والحدائق العامة وصناديق القمامة والمباني السكنية. أفادت الدراسات أن أكثر من 37 ألف طن

been reported that more than 37k tons and more than 24k tons are daily collected from the urban and rural sites, respectively (Bui et al. 2020). It almost contains 60% organic content in addition to other components such as plastic bags, papers, metal, cans, glass, and other little hazardous wastes including batteries, broken light bulbs, and electronic devices (Tran et al. 2020). While the industrial solid wastes are those which are collected from shops, hotels, markets, restaurants, which are also known as traditional solid wastes, in addition to the wastes excluding hazardous substances but collected from industries (Tran et al. 2020).

The agricultural solid wastes are derived from the processing of agricultural products such as fruits, crops, vegetables, in addition to the poultry, meat, and dairy products (Tran et al. 2020). Generally, the production and accumulation of municipal solid waste is resulted from multiple resources including the domestic, the industrial, and the commercial activates. It also includes the recyclable materials (such as glass,

وأكثر من 24 ألف طن يتم جمعها يوميًا من المواقع الحضرية والريفية، على التوالي (Bui et al. 2020). ويحتوي تقريبًا على محتوى عضوي بنسبة 60% بالإضافة إلى مكونات أخرى مثل الأكياس البلاستيكية والأوراق والمعادن والعلب والزجاج والمخلفات الخطرة الصغيرة الأخرى بما في ذلك البطاريات ومصابيح الإضاءة المكسورة والأجهزة الإلكترونية (Tran et al. 2020). في حين أن المخلفات الصناعية الصلبة هي تلك التي يتم جمعها من المتاجر والفنادق والأسواق والمطاعم، والتي تُعرف أيضًا باسم المخلفات الصلبة التقليدية، بالإضافة إلى المخلفات باستثناء المواد الخطرة ولكن يتم جمعها من الصناعات (Tran et al. 2020). تشتق المخلفات الزراعية الصلبة من معالجة المنتجات الزراعية مثل الفواكه والمحاصيل والخضروات، بالإضافة إلى الدواجن واللحوم ومنتجات الألبان (Tran et al. 2020). وبشكل عام، فإن إنتاج وتراكم النفايات البلدية الصلبة من موارد متعددة بما في ذلك الأنشطة المنزلية والصناعية والتجارية. كما تشمل المواد القابلة لإعادة التدوير (مثل الزجاج والمعادن والمواد البلاستيكية)، والمواد

metals, and plastics), the biodegradable materials (such as food remaining, fruit and vegetable peels), biomedical materials (such as disposable medical wastes and blood-contaminated cotton), and toxic materials (such as electronics, paints, and pesticides) (Taherzadeh et al. 2019).

6.3. Food waste and agricultural residues

Any raw, semi-processed, or processed substance that can be consumed by humans in addition to its inedible sections can be classified as food (Gustafsson et al. 2019). Spoiled food is a description for food that was deteriorated through some infrastructure restrictions in its life cycle such as manufacturing, processing, distribution, and storage. While food waste is a phase that describe the reduction in the amount or the quality of the food due to the actions of retailers, the providers of food services, and the behaviors of the customers (Mustafa and Maryam 2023). The global fruit production has been exceeded one billion tons in 2017 that resulted in huge amounts of wastes and by-products, as reported by the United Nations FAO. It has

القابلة للتحلل الحيوي (مثل بقايا الطعام، وقشر الفواكه والخضروات)، والمواد الطبية الحيوية (مثل المخلفات الطبية والقطن الملوث بالدم)، والمواد السامة (مثل الإلكترونيات والدهانات والمبيدات الحشرية) (Taherzadeh et al. 2019).

6.3. المخلفات الغذائية والبقايا الزراعية

يمكن تصنيف أي مادة خام أو شبه معالجة أو معالجة يمكن أن يستهلكها الإنسان بالإضافة إلى اجزائها الغير صالحة للأكل على أنها طعام (.Gustafsson et al 2019). الغذاء الفاسد هو وصف للغذاء الذي تدهور بسبب بعض قيود البنية التحتية في دورة حياته مثل التصنيع والمعالجة والتوزيع والتخزين. في حين أن إهدار الطعام هو مرحلة تصف الانخفاض في كمية أو جودة الطعام بسبب تصرفات تجار التجزئة ومقدمي الخدمات الغذائية وسلوكيات العملاء (Mustafa and Maryam 2023). وقد تجاوز إنتاج الفاكهة العالمي مليار طن في عام 2017 مما أدى إلى إنتاج كميات ضخمة من المخلفات والمنتجات الثانوية، كما ذكرت منظمة الأغذية والزراعة التابعة للأمم

been reported that 14.8% of Europe by-products were resulted as food wastes from fruit and vegetable industries (Fierascu et al. 2020). Fruits and vegetables wastes are known as the inedible sections of both of them that are wastes by farmers at the pre-consumer stage or through the food processing chain or later on by the consumers as the post-consumer chain.

On the other hand, most of the fruits and vegetables themselves are becoming wastes due to their short shelf life. Generally, huge amounts of these biomaterials are almost ended up in dump sites or occasionally burned (Kumar et al. 2020a). It has been reported that both of solid and liquid wastes are produced of food during the manufacturing, the preparation and consumption, and through the industrial processing. It is expected that, by the end of 2025, the worldwide food waste will be more than 2.2 billion tons (Wang et al. 2021).

Multiple waste management techniques of the food waste

المتحدة. أفادت التقارير أن 14.8% من المنتجات الثانوية في أوروبا نتجت عن مخلفات الطعام من صناعات الفاكهة والخضروات (Fierascu et al. 2020). تُعرف مخلفات الفواكه والخضروات بأنها الأجزاء غير الصالحة للأكل في كليهما والتي هي مخلفات من قبل المزارعين في مرحلة ما قبل الاستهلاك أو من خلال سلسلة تصنيع الأغذية أو في وقت لاحق من قبل المستهلكين باعتبارها سلسلة ما بعد الاستهلاك.

من ناحية أخرى، فإن معظم الفواكه والخضروات نفسها تصبح مخلفات بسبب فترة صلاحيتها القصيرة. وبصفة عامة، ينتهي الأمر بكميات هائلة من هذه المواد الحيوية تقريبًا في مواقع التفريغ أو أحيانًا يتم حرقها (Kumar et al. 2020a). وقد أفادت التقارير بأن كلاً من المخلفات الصلبة والسائلة يتم إنتاجها من الغذاء أثناء التصنيع والتحضير والاستهلاك ومن خلال المعالجة الصناعية. ومن المتوقع أن تكون مخلفات الطعام في جميع أنحاء العالم أكثر من 2.2 مليار طن بحلول نهاية عام 2025 (Wang et al. 2021).

تم تطبيق تقنيات متعددة لإدارة المخلفات الغذائية مثل المعالجات الحرارية

have been applied such as thermal treatments, extraction, slurry phase decomposition, and bioremediation. Among them all, the bioremediation technique is the most effective one, due to its minimal operational costs, low expenditures, and its harmfulness effects on the environment (Mustafa and Maryam 2023).

6.3.1. Bioremediation of domestic food waste

Domestic food waste is the waste that leftover uneaten after meals, as well as the partially consumed or unused food. This kind of waste is accounts of around 40% of the consumed food that will hence cost $28-33 for households each month. Insufficient national and international data about the food waste push the FAO to create the food loss index, where it calculates the amount of wasted food through the processes of manufacturing and supply before reaching the retail outlet (Van Geffen et al. 2020). According to this index, the FAO reported that 14% of food is lost before through the suppling process before reaching the consumer (Barrera and Hertel 2021).

والاستخلاص والتحليل والمعالجة الحيوية. ومن بين جميع هذه التقنيات، تُعد تقنية المعالجة الحيوية هي الأكثر فعالية، نظرًا لانخفاض تكاليفها التشغيلية، وانخفاض النفقات، وتأثيراتها الضارة على البيئة (Mustafa and Maryam 2023).

1.3.6. المعالجة الحيوية لمخلفات الطعام المنزلية

مخلفات الطعام المنزلية هي المخلفات التي لا يتم تناولها بعد الوجبات، بالإضافة إلى الطعام المستهلك جزئيًا أو غير المستخدم. يمثل هذا النوع من المخلفات حوالي 40% من الطعام المستهلك والذي سيكلف بالتالي 28-33 دولارًا للأسر كل شهر. إن عدم كفاية البيانات الوطنية والدولية حول هدر الغذاء يدفع منظمة الأغذية والزراعة إلى إنشاء مؤشر الفاقد من الغذاء، حيث يقوم بحساب كمية الطعام المهدر من خلال عمليات التصنيع والتوريد قبل الوصول إلى منفذ البيع بالتجزئة (.Van Geffen et al 2020). ووفقاً لهذا المؤشر، أفادت منظمة الأغذية والزراعة أن 14% من الغذاء يتم فقدانه من خلال عملية التوريد قبل وصوله إلى المستهلك (Barrera and Hertel

Household food waste is almost influenced by a lot of factors such as habits and emotions or bad food-related behavior. The average annual domestic waste for each consumer is almost 40-60% of food waste which contributes to about 20% in landfills. As a result, the United Nations has assigned a global goal in order to reduce the quantity of lost or wasted food by the year 2030 (Principato et al. 2021).

Scientists used various microorganisms for the treatment of food waste depending on many approaches such as the aerobic or an-aerobic digestion, liquefaction, composting, mechanical biological treatment, etc. For instance, the anaerobic digestion of domestic food waste is a process in which the microorganism can breakdown the organic substances in the absence of oxygen and can result in the production of energy (Mustafa and Maryam 2023). The domestic food waste has the advantage of having various kinds of nutrients such as lipids, carbohydrates, and proteins. On the other hand, composting is an extra method

2021). يتأثر هدر الطعام المنزلي بالكثير من العوامل مثل العادات والعواطف أو السلوك السيئ المتعلق بالطعام. ويبلغ متوسط النفايات المنزلية السنوية لكل مستهلك ما يقرب من 40-60% من النفايات الغذائية التي تساهم بحوالي 20% في مدافن النفايات. ونتيجة لذلك، حددت الأمم المتحدة هدفًا عالميًا لتقليل كمية الأغذية المفقودة أو المهدرة بحلول عام 2030 (Principato et al. 2021).

استخدم العلماء العديد من الكائنات الحية الدقيقة لمعالجة مخلفات الطعام اعتمادًا على العديد من الأساليب مثل الهضم الهوائي أو اللاهوائي، والتسييل، والتسميد، والمعالجة البيولوجية الميكانيكية، وما إلى ذلك. على سبيل المثال، الهضم اللاهوائي لمخلفات الطعام المنزلية هي عملية يعتمد فيها عل قدرة الكائنات الحية الدقيقة أن تحلل المواد العضوية في غياب الأكسجين ويمكن أن تؤدي إلى إنتاج الطاقة (Mustafa and Maryam 2023) . تتميز المخلفات الغذائية المنزلية بوجود أنواع مختلفة من العناصر الغذائية مثل الدهون والكربوهيدرات والبروتينات. ومن ناحية أخرى، يعد التسميد طريقة إضافية لمعالجة مخلفات الطعام المنزلية بواسطة الكائنات

for the treatment of domestic food waste by microorganisms and result in the degradation and stabilization of the organic matter in the waste (Rastogi et al. 2020). Household food waste has been also submitted for bioremediation process through the liquefaction process. It is a thermochemical process in which the biomass waste is treated through different chemical processes in a liquid media in order to have a liquefied product (Zhang et al. 2019a). The process can be accelerated by the addition of the microorganisms and their required nutrients into the treated biomass material. This process which is known as biological-liquefaction is a complexed but more successful process. However, the mechanical biological process is a combination of a mechanical step with another biological waste treatment step such as anaerobic digestion or composing (Mustafa and Maryam 2023).

الحية الدقيقة ويؤدي إلى تحلل وتثبيت المادة العضوية في المخلفات (Rastogi et al. 2020). كما تم أيضًا إخضاع مخلفات الطعام المنزلية لعملية المعالجة الحيوية من خلال عملية التسييل. إنها عملية كيميائية حرارية تتم فيها معالجة مخلفات الكتلة الحيوية من خلال عمليات كيميائية مختلفة في وسط سائل من أجل الحصول على منتج مسال (Zhang et al. 2019a). يمكن تسريع العملية عن طريق إضافة الكائنات الحية الدقيقة والمواد المغذية المطلوبة إلى مادة الكتلة الحيوية المعالجة. هذه العملية المعروفة باسم التسييل البيولوجي هي عملية معقدة ولكنها أكثر نجاحًا. إلا أن العملية البيولوجية الميكانيكية هي عبارة عن مزيج من خطوة ميكانيكية مع خطوة أخرى لمعالجة المخلفات البيولوجية مثل الهضم اللاهوائي أو التأليف (Mustafa and Maryam 2023).

6.3.2. Bioremediation of wastes from food processing industries

Every year, one-third of the worldwide food waste with the amount of 1.3 billion tons is produced as reported by FAO. The significant wastes originating through the food processed industries are from fruits and vegetables, dairy products, olive waste, meat, poultry and seafood by-products, and fermentation industries. According to the chemical composition of food waste, it is difficult to be recycled. The amount of food waste produced every year in a specific area is controlled by the population and their consumption patterns. In comparison between traditional food preparation techniques and processing food, it has been discovered that traditional techniques are resulting in very little wastes that almost used as feed, or composted, or disposed in the garbage bins. While the processing food including ready-to-eat meals are resulting in huge amounts of waste (JeyaSundar et al. 2020). Two methods of microbial bioremediation are applied for the industrial food waste. The first one, the

6.3.2. المعالجة الحيوية للمخلفات الناتجة عن الصناعات الغذائية

في كل عام، يتم إنتاج ثلث مخلفات الطعام في جميع أنحاء العالم بمقدار 1.3 مليار طن، وفقًا لما ذكرته منظمة الأغذية والزراعة. إن المخلفات الكبيرة الناتجة عن الصناعات الغذائية تأتي من الفواكه والخضروات ومنتجات الألبان ومخلفات الزيتون واللحوم والدواجن والمنتجات البحرية وصناعات التخمير. وفقا للتركيب الكيميائي للمخلفات الغذائية، فمن الصعب إعادة تدويرها. يتم التحكم في كمية مخلفات الطعام التي يتم إنتاجها كل عام في منطقة معينة من خلال السكان وأنماط استهلاكهم. بالمقارنة بين تقنيات تحضير الطعام التقليدية وتجهيز الطعام، فقد تم اكتشاف أن التقنيات التقليدية تنتج مخلفات قليلة جدًا تكاد تستخدم كعلف أو كسماد أو يتم التخلص منها في صناديق القمامة. في حين أن معالجة الأغذية بما في ذلك الوجبات الجاهزة للأكل تؤدي إلى كميات هائلة من النفايات (JeyaSundar et al. 2020). يتم تطبيق طريقتين للمعالجة الحيوية الميكروبية لمخلفات الأغذية الصناعية. الأول، الهضم اللاهوائي الميكروبي التقليدي، الذي يؤدي إلى إنتاج غاز الميثان.

traditional microbial anaerobic digestion, that result in the methane gas production. While the most recent one, is the using of microbial fuel cells or the microbial electrolysis cells. Bothe cells accelerate the bioconversion of complicated organic molecules of the industrial food waste into energy outputs (Mustafa and Maryam 2023).

6.3.3. Bioremediation of catering wastes

The remaining food of restaurants, kitchens, and catering facilities is known as the catering food. This kind of food is almost rich with processed fish and meat, uncooked vegetables, bakery products, cooking oil, and the peels of different fruits and vegetables. These catering wastes cause many environmental problems including for instance: high amounts of oily wastewater and oil fumes (Gao et al. 2019).

The food waste originating from restaurants has recently gotten a big attention all-over the world than the domestic food waste, as it is produced on a large scale and collection

في حين أن أحدثها هو استخدام خلايا الوقود الميكروبية أو خلايا التحليل الكهربائي الميكروبية. تعمل خلايا بوث على تسريع التحويل الحيوي للجزيئات العضوية المعقدة من مخلفات الاغذية الصناعية إلى مخرجات طاقة (Mustafa and Maryam 2023).

3.3.6. المعالجة الحيوية لمخلفات المطاعم

يُعرف الطعام المتبقي في المطاعم والمطابخ باسم الطعام المتخلف عن أماكن تحضير الطعام. ويكون هذا النوع من الطعام غنيًا ببقايا الأسماك واللحوم المصنعة، والخضروات غير المطهية، ومنتجات المخابز، وزيت الطهي، وقشور الفواكه والخضروات المختلفة. تتسبب مخلفات الطعام هذه في العديد من المشاكل البيئية بما في ذلك على سبيل المثال: كميات كبيرة من مياه الصرف الصحي النفطية وأبخرة النفط (Gao et al. 2019).

لقد حظيت مخلفات الطعام الناتجة عن المطاعم مؤخرًا باهتمام كبير في جميع أنحاء العالم أكثر من مخلفات الطعام

convenience. Recently, such food waste gain more academic, public, and environmental attention, as it might cause multiple environmental issues and cause danger to people's health (Yang et al. 2019). The anaerobic fermentation has been proposed as the best bioremediation technique for the treatment of catering wastes especially those containing lipids. Some microbial genera that were responsible for the degradation of amino acids, fatty acids, carbohydrates, and glycosaminoglycan have been included *Aspergillus, Pelomonas, Xeromyces, Methanobacterium, Faecalibacterium,* and *Corynebacterium.* The anaerobic co-digestion is another suggested bioremediation technique for the removal of vitamins, carbohydrates, and glycosaminoglycan by some microbial genera such as: *Methanobacterium, Fastidiosipila, Methanosaeta, Xeromyces,* and *Bifidobacterium* (Mustafa and Maryam 2023).

المنزلية، حيث يتم إنتاجها على نطاق واسع وسهولة التجميع. في الآونة الأخيرة، تحظى مخلفات الطعام هذه اهتمامًا أكاديميًا وعامًا وبيئيًا أكثر، لأنها قد تتسبب في العديد من القضايا البيئية وتتسبب في خطر على صحة الناس (Yang et al. 2019). تم اقتراح التخمير اللاهوائي كأفضل تقنية للمعالجة الحيوية لمعالجة مخلفات المطاعم وخاصة تلك التي تحتوي على الدهون. بعض الأجناس الميكروبية التي كانت مسؤولة عن تحلل الأحماض الأمينية، والأحماض الدهنية، والكربوهيدرات، والجليكوزامينوجليكان قد اشتملت على *Aspergillus, Pelomonas, Xeromyces, Methanobacterium, Faecalibacterium,* and *Corynebacterium.* الهضم اللاهوائي المشترك هو أسلوب آخر مقترح للمعالجة الحيوية لإزالة الفيتامينات والكربوهيدرات والجليكوزامينوجليكان بواسطة بعض الأجناس الميكروبية مثل: *Methanobacterium, Fastidiosipila, Methanosaeta, Xeromyces,* and *Bifidobacterium* (Mustafa and Maryam 2023).

6.4. Industrial waste

It includes all the inorganic or organic wastes that result from industry. They are classified into many categories according to different criteria.

They can be classified according to their nature into:

6.4.1. Organic industrial wastes

According to their chemical structures, they are organic in nature. They are resulted from wood manufactories, oil extraction, water treatment stations, food preservation, tanning, dyeing, painting, and plastic industries (khamis Soliman et al. 2019).

6.4.2. Inorganic industrial wastes

They are inorganic structures according to their chemistry in nature. They are released to the environment from different industries such as ceramic, cement, and granite factories (Amin et al. 2020).

They can be classified according to the characteristics of the pollution into:

6.4. المخلفات الصناعية

وتشمل جميع المخلفات العضوية أو غير العضوية الناتجة عن الصناعة. يتم تصنيفها إلى العديد من الفئات وفقا لمعايير مختلفة.

ويمكن تصنيفها حسب طبيعتها إلى:

6.4.1. المخلفات الصناعية العضوية

وفقا لبنيتها الكيميائية، فهي عضوية بطبيعتها. مثل الناتجة عن مصانع الأخشاب، واستخراج الزيوت، ومحطات معالجة المياه، وحفظ الأغذية، والدباغة، والصباغة، والدهان، والصناعات البلاستيكية (khamis Soliman et al. 2019).

6.4.2. المخلفات الصناعية الغير عضوية

وهي مركبات غير عضوية وذلك حسب تركيبها الكيميائي في الطبيعة. وتتواجد في البيئة من خلال صناعات مختلفة مثل مصانع السيراميك والأسمنت والجرانيت (Amin et al. 2020).

ويمكن تصنيفها حسب خصائص التلوث إلى:

6.4.3. Hazardous industrial waste

These industrial wastes are bearing hazardous properties. They are generating from industries such as plastic, rubber, and chemical industries (Amin et al. 2020; Soliman and Moustafa 2020).

6.4.4. Non-hazardous industrial waste

These industries bear non-hazardous properties and can be generated from clothing, food, and packaging industries (Amin et al. 2020; Soliman and Moustafa 2020).

They can be classified according to the industrial sectors into:

6.4.5. Mining industrial waste

These wastes are usually produced from mining industry including tailings and stones industry (Habib et al. 2020).

6.4.6. Metallurgical industrial waste

They are originating from the mining industries such as metals-related industries (Habib et al. 2020).

6.4.7. Chemical industrial waste

These wastes are released from medical-related industries such as pharmaceutical and medicine industries in addition to pesticide industries (Saravanan et al. 2022).

6.4.8. Food preservation industrial waste

These are the wastes that resulted from nutrition industries (Soliman and Moustafa 2020).

6.4.9. Constructional materials industrial waste

These are the wastes that originating from the building industries such as ceramic, cement, marble, steel, and granite industries (Soliman and Moustafa 2020).
They can be classified according to the industrial processes into:

6.4.6. مخلفات صناعة المعادن

وهي تنشأ من الصناعات التعدينية مثل الصناعات المرتبطة بالمعادن (Habib et al. 2020).

6.4.7. المخلفات الصناعية الكيميائية

وتنطلق هذه النفايات من الصناعات المرتبطة بالكيماويات مثل الصناعات الدوائية والأدوية بالإضافة إلى صناعات المبيدات الحشرية (Saravanan et al. 2022).

6.4.8. المخلفات الصناعية من حفظ الاغذية

وهي المخلفات الناتجة عن الصناعات الغذائية (Soliman and Moustafa 2020).

6.4.9. المخلفات الصناعية من مواد البناء

وهي المخلفات الناتجة من صناعة مواد البناء مثل صناعات السيراميك والأسمنت والرخام والصلب والجرانيت (Soliman and Moustafa 2020).

ويمكن تصنيفها حسب العمليات الصناعية إلى:

6.4.10. Fired industrial waste

These wastes are produced from the fire-based industries such as ceramic, bricks, and steel industries (Amin et al. 2020).

6.4.11. Unfired industrial waste

These wastes are resulted from the unfired-based industries such as paper, food preservative, and marble industries (khamis Soliman et al. 2019).

6.5. Medical waste

Medical wastes can be classified into dangerous, non-hazardous, and infectious wastes. According to the world health organization, about 85% of the medical wastes are non-hazardous wastes compared with the remaining 15% which are considered as hazardous substances that might be infectious, chemicals, radioactive, or dangerous (Chandra et al. 2023). If not properly disposed of or treated; these wastes will cause extreme impacts on the human health and the surrounding environment. In some cases, the open burning

---

6.4.10. حرق المخلفات الصناعية

تنتج هذه المخلفات من الصناعات المعتمدة على الحرائق مثل صناعة السيراميك والطوب والصلب (Amin et al. 2020).

6.4.11. المخلفات الصناعية غير المشتعلة

وتنتج هذه المخلفات من الصناعات الغير مشتعلة مثل صناعة الورق والمواد الحافظة الغذائية والرخام (khamis Soliman et al. 2019).

6.5. المخلفات الطبية

ويمكن تصنيف المخلفات الطبية إلى مخلفات خطرة، وغير خطرة، ومعدية. ووفقا لمنظمة الصحة العالمية فإن حوالي 85% من المخلفات الطبية هي مخلفات غير خطرة مقارنة مع 15% المتبقية والتي تعتبر مواد خطرة قد تكون معدية أو كيميائية أو مشعة أو خطيرة (Chandra et al. 2023). إذا لم يتم التخلص من هذه المخلفات أو علاجها بشكل صحيح؛ فانها تتسبب في آثار خطيرة على صحة الإنسان والبيئة المحيطة. وفي بعض الحالات، يكون الحرق في الهواء الطلق من أكثر الطرق شيوعًا لمعالجة المخلفات الطبية.

and incinerations are the most common methods of the treatment of medical wastes. However, these treatment methods can result in the formation of dioxins, furans, and other materials that considered carcinogenic and common cause for respiratory diseases. One another problem is the improper disposal of sharp medical wastes such as needles, blades, syringes, and scalpels that lead to a usual injuries and/or infections to the health-workers and/or the waste-handlers. In addition, the improper disposal of pharmaceutical wastes such as contaminated drugs, vaccines, and expired drugs can largely pollute the soil or the water bodies and badly affecting the crops or the aquatic life (Ye et al. 2022). Figure 8 is describing the amount of medical wastes generated by different countries worldwide.

ومع ذلك، فإن طرق العلاج هذه يمكن أن تؤدي إلى تكوين الديوكسينات والفيورانات وغيرها من المواد التي تعتبر مسرطنة وسبب شائع لأمراض الجهاز التنفسي. إحدى المشاكل الأخرى هي التخلص غير السليم من المخلفات الطبية الحادة مثل الإبر والشفرات والمحاقن والمباضع التي تؤدي إلى إصابات و/أو عدوى معتادة للعاملين الصحيين و/أو من يتعاملون مع المخلفات. بالإضافة إلى ذلك، فإن التخلص غير السليم من المخلفات الصيدلانية مثل الأدوية الملوثة واللقاحات والأدوية منتهية الصلاحية يمكن أن يلوث التربة أو المسطحات المائية إلى حد كبير ويؤثر بشكل سيء على المحاصيل أو الحياة المائية (Ye et al. 2022). ويصف الشكل 8 كمية النفايات الطبية الناتجة عن مختلف دول العالم.

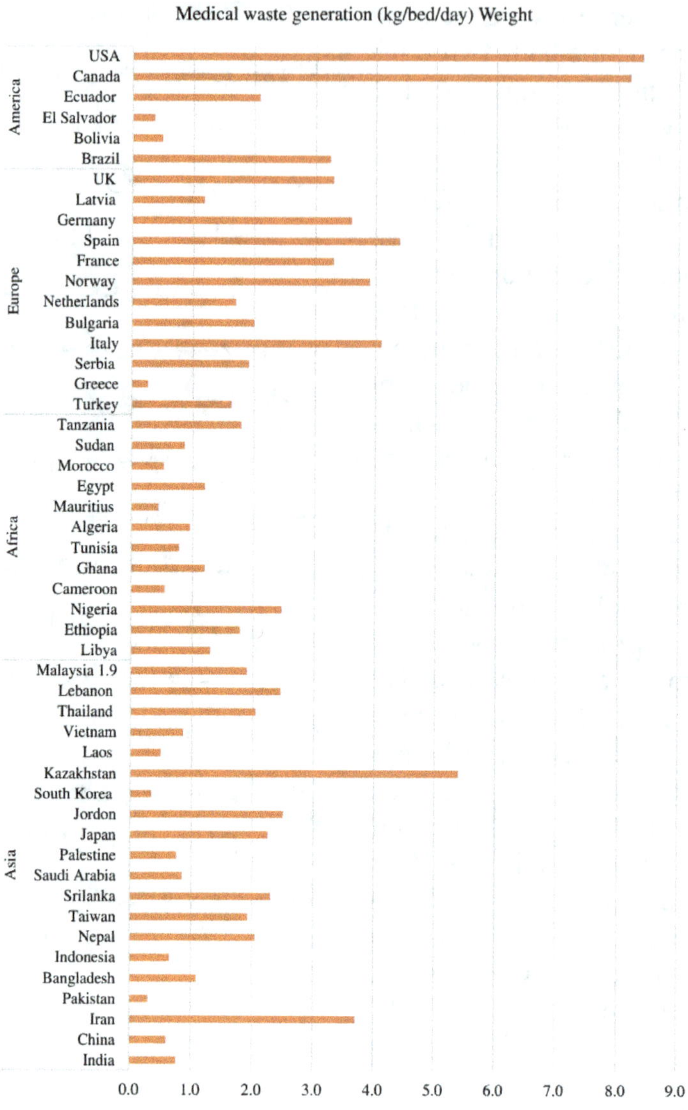

Figure 8 An estimation of the average comparable amounts of the medical wastes produced by the top 50 countries (kg/bed/day) (Chandra et al. 2023). الشكل 8 تقدير لمتوسط الكميات القابلة للمقارنة من المخلفات الطبية التي تنتجها أكبر 50 دولة (كجم/سرير/يوم) (Chandra et al. 2023).

6.5.1. The environmental impact of medical wastes

In recent years, and due to the advances in the medical instruments and the due to the emerging of pandemic and epidemic diseases, the amount and type of medical wastes are continuously increasing, which needs a lot of research and development in the most probable techniques for the removal and getting rid of medical wastes (Torres Munguía et al. 2022), in addition to the regular updating of the guidelines related to the removal and practices of medical wastes. As the improper control of such wastes can result in a lot of environmental damage and health risks for animals and humans (Manzoor and Sharma 2019).

According to (Lenzen et al. 2020; Wei et al. 2021), the releasing of medical wastes into the environment can cause bio harms such as:

- **Air pollution :** the open burning of medical wastes can result in the release of hazardous chemicals including furans, dioxins, and mercury which can cause air pollution through their leakage to the

1.6.5.1. التأثير البيئي للمخلفات الطبية

في السنوات الأخيرة، ونظراً لتقدم الأجهزة الطبية وظهور الأمراض الوبائية، فإن كمية ونوعية المخلفات الطبية في تزايد مستمر، الأمر الذي يحتاج إلى الكثير من البحث والتطوير في التقنيات الأكثر احتمالاً للتخلص وإزالة المخلفات الطبية (Torres Munguía et al. 2022)، بالإضافة إلى التحديث المنتظم للمبادئ التوجيهية المتعلقة بإزالة المخلفات الطبية وممارساتها. حيث أن التحكم غير السليم في مثل هذه المخلفات يمكن أن يؤدي إلى الكثير من الأضرار البيئية والمخاطر الصحية على الحيوان والإنسان (Manzoor and Sharma 2019).

وطبقاً لـ (Lenzen et al. 2020; Wei et al. 2021) فان إطلاق المخلفات الطبية في البيئة يمكن أن يسبب أضرارًا حيوية مثل:

- **تلوث الهواء**: يمكن أن يؤدي حرق المخلفات الطبية في الهواء الطلق إلى إطلاق مواد كيميائية خطرة بما في ذلك الفوران والديوكسينات والزئبق والتي يمكن أن تسبب تلوث الهواء من خلال تسربها إلى

atmosphere (Manisalidis et al. 2020).

- **Water pollution:** the medical waste that are not properly disposed of can easily contaminate the water bodies and causing a lot of health risks to all wildlife as well as the humans.

- **Soil pollution:** the untreated released medical wastes can strongly contaminate the soil and the landfills and resulting in severe danger to the human health and to environment (Manzoor and Sharma 2019).

- **Transmission of diseases:** some medical waste might include parasites, viruses, and pathogenic microbes that can spread illness to the contacted animals or the humans (Das et al. 2021; Rajak et al. 2022).

6.5.2. Biotechnological managing of medical wastes

Biotechnology is one of the most promising applications that can be applied for management of different types of wastes such as infectious, radioactive, and pharmaceutical wastes (Figure 9). Such technology can entirely prevent the release of

الغلاف الجوي (Manisalidis et al. 2020).

ـ **تلوث المياه**: المخلفات الطبية التي لا يتم التخلص منها بشكل صحيح يمكن أن تلوث المسطحات المائية بسهولة وتسبب الكثير من المخاطر الصحية على الحياة البرية والإنسان.

ـ **تلوث التربة**: يمكن أن تؤدي المخلفات الطبية المنطلقة غير المعالجة إلى تلويث التربة بشدة ويؤدي إلى خطر شديد على صحة الإنسان والبيئة (Manzoor and Sharma 2019).

ـ **نقل الأمراض**: قد تحتوي بعض المخلفات الطبية على طفيليات وفيروسات وميكروبات ممرضة يمكن أن تنشر المرض إلى الحيوانات أو البشر (Das et al. 2021; Rajak et al. 2022).

6.5.2. الإدارة التكنولوجية الحيوية للمخلفات الطبية

تعد التكنولوجيا الحيوية واحدة من أكثر التطبيقات الواعدة التي يمكن تطبيقها لإدارة أنواع مختلفة من المخلفات مثل المخلفات المعدية والمشعة والنفايات الصيدلانية (الشكل 9). يمكن لهذه التكنولوجيا أن تمنع تمامًا إطلاق المركبات الضارة مثل

harmful compounds such as furans, heavy metals, and dioxins into the environment (Parmieka, 2020). In a recent study, the researches proposed important factors for the reduction of medical wastes such as: raising the people awareness, training on the proper management of medical wastes, and training on the proper procedures for medical waste management (Lee and Lee 2022). The following procedures can comprehensively explain the key factors for medical waste management:

6.5.2.1. Bioremediation of medical wastes

The advanced bioremediation of the medical wastes is hugely depending on the using of recombinant enzymes. These enzymes have been genetically modified in order to have more sensible activity and stability, which help them to perfectly eliminate the toxins or harms of the medical wastes (Humer et al. 2020). For instance, the recombinant enzyme named horseradish peroxidase (HRP), which contains the heme group, has been used to catalyze the phenols oxidation

الفوران والمعادن الثقيلة والديوكسينات في البيئة (Parmieka, 2020). وفي دراسة حديثة اقترحت الأبحاث عوامل مهمة للحد من المخلفات الطبية مثل: رفع وعي الناس، والتدريب على الإدارة السليمة للمخلفات الطبية، والتدريب على الإجراءات الصحيحة لإدارة المخلفات الطبية (Lee and Lee 2022). يمكن للإجراءات التالية أن تشرح بشكل شامل العوامل الرئيسية لإدارة المخلفات الطبية:

6.5.2.1. المعالجة الحيوية للمخلفات الطبية

تعتمد المعالجة الحيوية المتقدمة للمخلفات الطبية بشكل كبير على استخدام الإنزيمات المؤتلفة. وقد تم تعديل هذه الإنزيمات وراثيا لكي يكون لها نشاط واستقرار أكثر معقولية، مما يساعدها على التخلص تماما من السموم أو أضرار المخلفات الطبية (Humer et al. 2020). على سبيل المثال، تم استخدام الإنزيم المؤتلف المسمى بيروكسيداز الفجل (HRP)، والذي يحتوي على مجموعة الهيم، لتحفيز أكسدة الفينولات والهيدروكربونات العطرية

and polycyclic aromatic hydrocarbons, which are frequent components of the medical waste. This recombinant enzyme can be effectively used for the bioremediation of medical wastes through its immobilization on a solid support (Ahmed et al. 2022). Laccase enzyme is another example for the recombinant enzymes that can be bioremediate the environmental wastes. The enzyme contains copper element and has the ability to oxidize the lignin molecule, which is a complex polymer extracted form plants and considered a significant part of the medical waste streams (Singh and Gupta 2020). Similarly, the combination of recombinant forms of the proteases, lipases, and cellulase enzymes can accelerate the breakdown of proteins, lipids, and cellulose parts of the medical wastes (Yadav et al. 2022).

متعددة الحلقات، والتي تعد مكونات متكررة للمخلفات الطبية. يمكن استخدام هذا الإنزيم المؤتلف بشكل فعال في المعالجة الحيوية للمخلفات الطبية من خلال تثبيته على دعامة صلبة (Ahmed et al. 2022). يعد إنزيم لاكيز مثالًا آخر على الإنزيمات المؤتلفة التي يمكن أن تعالج المخلفات البيئية بيولوجيًا. يحتوي الإنزيم على عنصر النحاس وله القدرة على أكسدة جزيء اللجنين وهو عبارة عن بوليمر معقد يستخرج من النباتات ويعتبر جزءا هاما فى معالجة المخلفات الطبية (Singh and Gupta 2020). وبالمثل، فإن الجمع بين الأشكال المؤتلفة من البروتياز والليباز وإنزيمات السليلوز يمكن أن يؤدي إلى تسريع تحلل البروتينات والدهون وأجزاء السليلوز من المخلفات الطبية (Yadav et al. 2022).

6.5.2.2. Composting of medical wastes

The composting of medical wastes is another natural and sustainable management method. In this method, the biodegradation of the organic molecules of the medical waste can help for the creation of a lot of nutrients in the soil that could be used for agriculture and gardening applications. It would be efficient to compost a variety of medical wastes such as food wastes of the hospitals or the expiry dated drugs or other organic matters (Li et al. 2023). The composting process is the work of different microorganisms such as fungi, actinomycetes, and bacteria. These microorganisms are degrading the organic matter of the waste into carbon dioxide and water in addition to some organic acids. The production of heat during this process is an advantage, as it causes the killing of germs that might be existed in the waste (Sufficiency et al. 2022). One of the most used soil bacteria for the composting process is *Bacillus subtilis*. It has been successfully applied for the degradation of wide range of organic molecules and resulted in high-quality

2.2.5.6. تسميد المخلفات الطبية

يعد تحويل المخلفات الطبية إلى سماد طريقة أخرى للإدارة الطبيعية والمستدامة. في هذه الطريقة، يمكن أن يساعد التحلل الحيوي للجزيئات العضوية للمخلفات الطبية في خلق الكثير من العناصر الغذائية في التربة التي يمكن استخدامها في تطبيقات الزراعة. سيكون من الفعال تحويل مجموعة متنوعة من المخلفات الطبية إلى سماد مثل مخلفات الطعام في المستشفيات أو الأدوية التي انتهت صلاحيتها أو المواد العضوية الأخرى (Li et al. 2023). عملية التسميد هي عمل الكائنات الحية الدقيقة المختلفة مثل الفطريات، والفطريات المشعه، والبكتيريا. وتقوم هذه الكائنات الحية الدقيقة بتحليل المادة العضوية لمخلفات إلى ثاني أكسيد الكربون وماء بالإضافة إلى بعض الأحماض العضوية. ويعد إنتاج الحرارة خلال هذه العملية ميزة لأنها تتسبب في قتل الجراثيم التي قد تكون موجودة في النفايات (Sufficiency et al. 2022). هناك نوع من بكتيريا التربة الأكثر استخدامًا في عملية التسميد هي *Bacillus subtilis* وقد تم تطبيقها بنجاح لتحلل مجموعة

compost (Duan et al. 2020; Mahapatra et al. 2022). Similarly, *Aspergillus niger*, is a fungus that has been successfully applied as important step in the composing process. It has the capability to produce enzymes required for the breakdown of pharmaceutical and food wastes. It can effectively degrade the lignin and cellulose structures into less complexed materials that can be easily degraded by the bacterial cells existed in the compost pile (Chandra et al. 2023).

Other enzymes such as proteases, cellulases, and lipases can be also used for the biodegradation of different substrates in the medical waste. Cellulases can be used for the degradation of multiple medical waste ingredients such as gauzes, cotton swabs, and bandages. The advantages of such degradation are reducing of the waste volume in addition to the production of nutrient-rich compost. Lipases and proteases can be also used for the biodegradation of lipids/fats and proteins in the medical waste, respectively. Such degradation outcomes can help for lowering the risk

واسعة من الجزيئات العضوية إلى سماد عالي الجودة (Duan et al. 2020; Mahapatra et al. 2022). وبالمثل، يعتبر فطر *Aspergillus niger* من الفطريات التي تم تطبيقها بنجاح كخطوة مهمة في عملية التركيب حيث لديها القدرة على إنتاج الإنزيمات اللازمة لتفكيك المخلفات الدوائية والغذائية. وايضا يمكنه تحليل هياكل اللجنين والسليلوز بشكل فعال إلى مواد أقل تعقيدًا يمكن أن تتحلل بسهولة بواسطة الخلايا البكتيرية الموجودة في كومة السماد (Chandra et al. 2023).

يمكن أيضًا استخدام إنزيمات أخرى مثل البروتياز والسيلولاز والليباز في التحلل الحيوي للركائز المختلفة في المخلفات الطبية. يمكن استخدام السليوليز لتحليل مكونات المخلفات الطبية المتعددة مثل الشاش، ومسحات القطن، والضمادات. تتمثل مزايا هذا التحليل في تقليل حجم المخلفات بالإضافة إلى إنتاج السماد الغني بالمغذيات. يمكن أيضًا استخدام الليباز والبروتياز في التحلل الحيوي للدهون/الدهون والبروتينات الموجودة في المخلفات الطبية، على التوالي. يمكن أن تساعد نتائج التحلل هذه في تقليل خطر

of infection, lowering the environmental risks and also valorizing the obtained compost (Patel et al. 2019). The composting of the medical wastes is an effective alternative to landfills dumping. It creates a fertile and healthy soil that can be effectively used for agriculture purposes, instead of the discharging of dangerous toxins from these wastes into the environment (Chandra et al. 2023).

6.5.2.3. Phytoremediation of medical wastes

It is a bioremediation process in which the plants and their combining microorganisms can remove or detoxify the environmental pollutants. It has shown a promising activity for the bioremediation of chemicals, germs, and medical wastes (Ansari et al. 2020). The phytoremediation process can provide a safe, long-lasting, and affordable bioremediation alternative. *Chrysopogon zizanioides* which is known as vetiver grass has been successfully applied for the removal of organic pollutants, pathogens, and heavy metals from medical wastes. Its rooting system has a potential capability to absorb and store multiple pollutants inside its tissues with the advantage of delaying their release into the environment (Liu and Tran 2021). Similarly, willow plant which is scientifically known as *Salix* sp. can effectively store the toxins of the medical waste inside its tissues, in addition to its ability to encourage the propagation of the beneficial soil microbes that can speed up the bioremediation and detoxification process (Landberg and Greger 2022).

الطبية. يتمتع نظام المعالجة النباتية بقدرة على امتصاص وتخزين العديد من الملوثات داخل أنسجتها مع ميزة تأخير إطلاقها في البيئة (Liu and Tran 2021). وكذلك نبات الصفصاف والذي يعرف علمياً باسم *Salix* sp يمكن أن يخزن بشكل فعال سموم المخلفات الطبية داخل أنسجته، بالإضافة إلى قدرته على تشجيع تكاثر ميكروبات التربة المفيدة التي يمكن أن تسرع عملية المعالجة الحيوية وإزالة السموم (Landberg and Greger 2022).

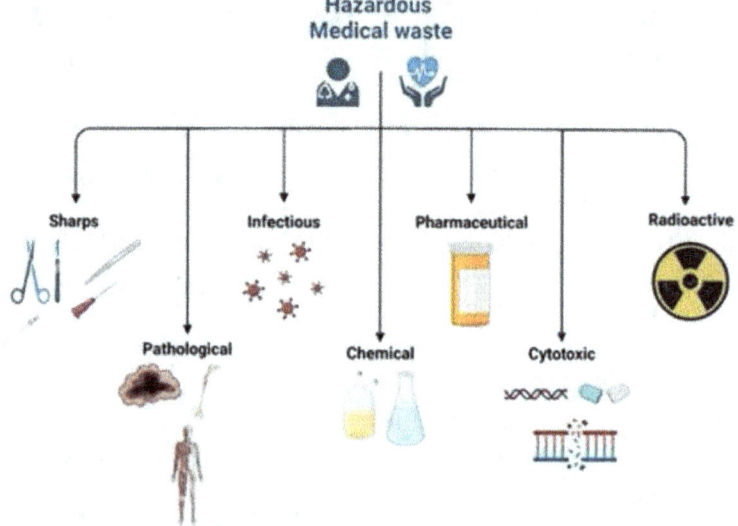

Figure 9 Types of hazardous medical wastes (Attrah et al. 2022).

الشكل 9 أنواع المخلفات الطبية الخطرة (Attrah et al. 2022).

6.6. Electronic wastes (e-waste)

Since the beginning of 20th century, a remarkable progress in the fields of engineering, science, and technology has been developed. This recent science revolution resulted in the creation of amazing tools and devices such as electrical and electronic devices that become normal parts of our daily lives. Although this technological revolution aimed to elevate our living standards and enhancing our life styles, bur causing two environmental distressing impacts which are the pollution and the reduction in the natural resources (Han et al. 2022).

The reports of the Global E-waste Monitor 2020, recorded the release of 53.6 million metric tons of electronic and electrical waste in 2019. The largest portion of these wastes were the small electronic devices followed by the large instruments (Figure 10) (Han et al. 2022; Forti et al. 2020).

It has been reported that e-waste is fastest growing waste stream with annual expected growth rate of 3-5%. It is

6.6. المخلفات الإلكترونية (النفايات الإلكترونية)

منذ بداية القرن العشرين، حدث تقدم ملحوظ في مجالات الهندسة والعلوم والتكنولوجيا. أدت هذه الثورة العلمية الحديثة إلى إنشاء أدوات وأجهزة مذهلة مثل الأجهزة الكهربائية والإلكترونية التي أصبحت جزءًا طبيعيًا من حياتنا اليومية. وعلى الرغم من أن هذه الثورة التكنولوجية تهدف إلى رفع مستويات معيشتنا وتعزيز أنماط حياتنا، إلا أنها تسبب أثرين بيئيين مزعجين هما التلوث وانخفاض الموارد الطبيعية (Han et al. 2022).

سجلت تقارير المرصد العالمي للمخلفات الإلكترونية 2020 إطلاق 53.6 مليون طن متري من المخلفات الإلكترونية والكهربائية في عام 2019. وكان الجزء الأكبر من هذه المخلفات عبارة عن الأجهزة الإلكترونية الصغيرة تليها الأدوات الكبيرة (الشكل 10) (Han et al. 2022; Forti et al. 2020). تشير التقارير إلى أن المخلفات الإلكترونية هي أسرع المخلفات نموًا بمعدل نمو سنوي متوقع يتراوح بين 3-5%. ومن المتوقع أن

يصل حجم المخلفات الإلكترونية عالميًا إلى 74 مليون طن بحلول نهاية عام 2030 (Forti et al. 2020).

expected that the e-waste will be globally reached up to 74 million tons by end of 2030 (Forti et al. 2020).

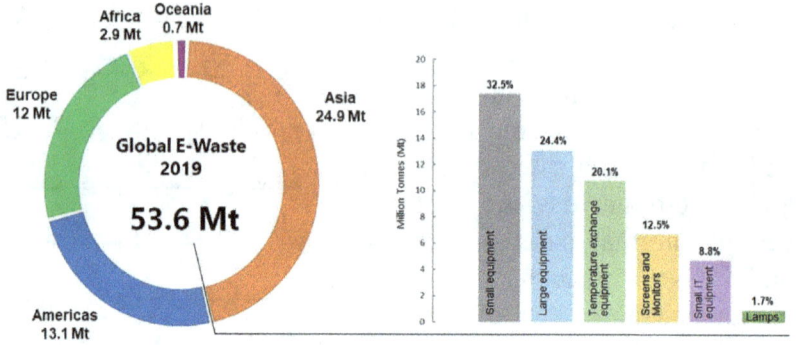

Figure 10 The global e-waste released in 2019 as reported by the Global E-waste Monitor 2020 (Han et al. 2022).

الشكل 10 المخلفات الإلكترونية العالمية الصادرة في عام 2019 وفقًا لما ورد في المرصد العالمي للمخلفات الإلكترونية 2020 (Han et al. 2022).

تحتوي المخلفات الإلكترونية على مكونات مختلفة حسب فئتها أو خطوط إنتاجها. وقد تحتوي على معادن ثمينة ومواد أرضية نادرة والسيراميك والبلاستيك، بالإضافة إلى معادن خطرة مثل الرصاص والكروم والنيكل والزئبق والملوثات العضوية بما في ذلك ثنائي الفينيل متعدد الكلور أو مثبطات اللهب المبرومة (.Singh et al 2020a). يمكن أن يؤدي إطلاق مثل هذه الملوثات في البيئة إلى توليد الكثير من المكونات السامة التي يمكن أن تشكل خطورة على الإنسان والبيئة (Cesaro et

The e-waste is containing different constituents according to their category or product lines. They may have precious metals, rare earth materials, ceramics, and plastics, in addition to hazardous metals such as lead, chromium, nickel, mercury, and organic pollutants including polychlorinated biphenyls or brominated flame retardants (Singh et al. 2020a). The releasing of such pollutants into the environment can generating a lot of toxic components that can be dangerous for the human as well as the environment (Cesaro et al. 2019; Rautela et al. 2021). The management of the e-waste is a big challenge that should be handled. These e-wastes can be used as alternative and secondary source for the raw materials for the extraction of precious elements that already existed in the electronic devices such as silver, gold, ruthenium, palladium, and platinum, as this principle the totally matched with principles of circular economy (Han et al. 2022).

6.6.1. Biotechnological recycling of e-waste

Both of chemical, mechanical, and physical methods have been applied for the recovery of valuable substances from e-waste (Rawat et al. 2020). The emerging technologies are aiming to the transform of e-waste components into high value products. For instance, plastics which are almost 20% of e-waste can be extracted and subjected for recycling. On the other hand, the metals of e-waste can be recovered using various processes such as pyrometallurgical or hydrometallurgical procedures. The

(al. 2019; Rautela et al. 2021).
تمثل إدارة النفايات الإلكترونية تحديًا كبيرًا يجب التعامل معه. ويمكن استخدام هذه المخلفات الإلكترونية كمصدر بديل وثانوي للمواد الخام اللازمة لاستخلاص العناصر الثمينة الموجودة بالفعل في الأجهزة الإلكترونية مثل الفضة والذهب والروثينيوم والبلاديوم والبلاتين، حيث يتوافق هذا المبدأ تمامًا مع المبادئ الاقتصاد الدائري (Han et al. 2022) .

6.6.1. إعادة تدوير المخلفات الإلكترونية باستخدام التكنولوجيا الحيوية

تم تطبيق كل من الطرق الكيميائية والميكانيكية والفيزيائية لاستعادة المواد القيمة من المخلفات الإلكترونية (Rawat et al. 2020). تهدف التقنيات الناشئة إلى تحويل مكونات المخلفات الإلكترونية إلى منتجات عالية القيمة. على سبيل المثال، يمكن استخراج المواد البلاستيكية التي تشكل ما يقرب من 20% من المخلفات الإلكترونية وإخضاعها لإعادة التدوير. ومن ناحية أخرى، يمكن استعادة معادن المخلفات الإلكترونية باستخدام عمليات مختلفة مثل إجراءات التعدين الحراري أو

pyrometallurgical procedure depending on various treatments such as alkali chemical treatment or furnace smelting, thermal treatment, and gas/liquid/solid reactions at elevated temperatures. While the hydrometallurgical treatments include leaching, adsorption, ion exchange, or solvent extraction processes (Ahirwar and Tripathi 2021). These traditional procedures have multiple disadvantages such as the releasing of toxic chemicals, cyanide, or dioxins. However, in order to decrease the amount of released toxic pollutants, the recent treatments such as electrochemical and vacuum metallurgical technologies have been developed alone, and sometimes combined with the enhanced traditional methods.

The microbial-assisted treatments of e-waste have been evolved as alternative green methods for the already applied conventional methods. Some microbes can naturally take the metals from the surrounding environment in order to be used for their structural and catalytic functions. These microbes that are almost belonging to the

التعدين المائي. يعتمد إجراء المعالجة المعدنية الحرارية على معالجات مختلفة مثل المعالجة الكيميائية القلوية أو صهر الفرن، والمعالجة الحرارية، وتفاعلات الغاز/السائل/الصلبة في درجات حرارة مرتفعة. بينما تشمل المعالجات الميتالورجية المائية عمليات الترشيح أو الامتزاز أو التبادل الأيوني أو استخلاص المذيبات (Ahirwar and Tripathi 2021). هذه الإجراءات التقليدية لها عيوب متعددة مثل إطلاق المواد الكيميائية السامة أو السيانيد أو الديوكسينات. ومع ذلك، من أجل تقليل كمية الملوثات السامة المنبعثة، تم تطوير المعالجات الحديثة مثل التقنيات الكهروكيميائية والتعدين الفراغي بمفردها، وفي بعض الأحيان يتم دمجها مع الطرق التقليدية المحسنة.

لقد تم تطوير المعالجات بمساعدة الميكروبات للمخلفات الإلكترونية كطرق خضراء بديلة للطرق التقليدية المطبقة بالفعل. يمكن لبعض الميكروبات أن تأخذ المعادن بشكل طبيعي من البيئة المحيطة لاستخدامها في وظائفها الهيكلية والتحفيزية. يمكن لهذه الميكروبات التي تنتمي تقريبًا إلى الأنواع الميكروبية ذات

chemolithotrophic and heterotrophic microbial species, can naturally convert the insoluble metals into soluble and extractable forms through the direct action that involves the surface attachment of the metal following by oxidizing the minerals and converting the metals into soluble form, or indirect action that involves the generation of ferric ions that will act as the oxidizing agents (Figure 11). This bioleaching process can be performed through various mechanisms of acidolysis, redoxolysis, or bioaccumulation (Srichandan et al. 2019).

Multiple bacterial genera such as *Pseudomonas* sp. and *Acidithiobacillus* sp. and fungal genera such as *Aspergillus* sp. have been shown a potent capacity for the recovery of metals such as Zn, Fe, Ni, and Cu and precious metals such as Pt, Au, and Ag from e-waste. The bioleaching of e-waste has been particularly performed by the acidophilic bacteria as they have tolerance to heavy metals (Mudila et al. 2021; Zhao and Wang 2019).

التغذية الكيميائية والمغايرة أن تحول المعادن غير القابلة للذوبان بشكل طبيعي إلى أشكال قابلة للذوبان وقابلة للاستخلاص من خلال العمل المباشر الذي يتضمن الارتباط السطحي للمعدن التالي عن طريق أكسدة المعادن وتحويل المعادن إلى شكل قابل للذوبان، أو بشكل غير مباشر الإجراء الذي يتضمن توليد أيونات الحديديك التي ستكون بمثابة العوامل المؤكسدة (الشكل 11). يمكن إجراء عملية التصفية الحيوية هذه من خلال آليات مختلفة للتحلل الحمضي، أو تحلل الأكسدة، أو التراكم الحيوي (Srichandan et al. 2019).

أجناس بكتيرية متعددة مثل *Pseudomonas* sp. and *Acidithiobacillus* sp والأجناس الفطرية مثل *Aspergillus* sp. لقد أظهرت قدرة قوية على استعادة المعادن مثل Zn وFe وNi وCu والمعادن الثمينة مثل Pt وAu وAg من المخلفات الإلكترونية. وقد تم إجراء التصفية الحيوية للمخلفات الإلكترونية بشكل خاص بواسطة البكتيريا المحبة للحموضة نظرًا لقدرتها على تحمل المعادن الثقيلة (Mudila et al. 2021; Zhao and Wang 2019).

Printed circuit boards (PCBs) are among the targets of recycling as they have more than 95% of metals that can be recovered. They are composed of multiple metals that accounts for 20-30%. Approximate percentage of 20% of them are copper and approximate amount of 0.3-0.4% are precious metals such as platinum, gold, and silver. The component heterogeneity and toxicity of the PCBs are the main reasons that negatively affecting the microbial growth resulting in low efficiency and long leaching time. A lot of research efforts have been dedicated for the increasing of the efficiency of the bioleaching process. Recent researches are interested in improving the tolerance of the microbial cells against the e-waste toxicity through the preadaptation and acclimatization of the cells to gradual PCBs concentrations before the bioleaching process. A further step in order to increase the bioleaching activity was followed through an indirect and non-contact mechanism in which the bacterial-free supernatant was used instead of the cells itself (Sadeghabad et al. 2019).

تعد لوحات الدوائر المطبوعة (PCBs) من بين أهداف إعادة التدوير حيث أنها تحتوي على أكثر من 95% من المعادن التي يمكن استردادها. وهي مكونة من معادن متعددة تمثل 20-30%. نسبة تقريبية 20% منها عبارة عن نحاس ونسبة تقريبية 0.3-0.4% عبارة عن معادن ثمينة مثل البلاتين والذهب والفضة. يعد عدم تجانس المكونات وسمية مركبات ثنائي الفينيل متعدد الكلور من الأسباب الرئيسية التي تؤثر سلبًا على نمو الميكروبات مما يؤدي إلى انخفاض الكفاءة ووقت الترشيح الطويل. لقد تم تخصيص الكثير من الجهود البحثية لزيادة كفاءة عملية الترشيح الحيوي. تهتم الأبحاث الحديثة بتحسين قدرة تحمل الخلايا الميكروبية ضد سمية النفايات الإلكترونية من خلال التكيف المسبق للخلايا وتأقلمها مع التركيزات التدريجية لمركبات ثنائي الفينيل متعدد الكلور قبل عملية التصفية الحيوية. تم اتباع خطوة أخرى من أجل زيادة نشاط التصفية الحيوية من خلال آلية غير مباشرة وغير متصلة حيث تم استخدام المادة الطافية الخالية من البكتيريا بدلاً من الخلايا نفسها (Sadeghabad et al. 2019). هناك عوامل أخرى يمكن أن تؤثر بشكل إيجابي

Another factors that can positively affect the bioleaching process are the using of multiple microbial species as consortia in addition to the optimization of the abiotic factors such as pH, temperature, carbon source, and pulp density (Han et al. 2022).

6.6.2. Potential biodegradation of e-waste plastics

The plastic portion of the e-waste is approximately 37%. These plastics are nearly including 15 different types of polymers with the following composition: 27% polystyrene, 24% acrylonitrile, 7% polycarbonate, 5% polypropylene, 2% polyethylene, and other thermoplastics and polymer blends that form the remaining part. Some of these plastics can be subjected to recycling; however, their biodegradation can offer eco-friendlier strategy. Several microorganisms have been reported as plastic degraders (Ru et al. 2020; Ali et al. 2021a). Various microorganisms that were isolated from different locations including landfill

على عملية التصفية الحيوية وهي استخدام الأنواع الميكروبية المتعددة كاتحادات بالإضافة إلى تحسين العوامل اللاأحيائية مثل الرقم الهيدروجيني ودرجة الحرارة ومصدر الكربون وكثافة اللب (Han et al. 2022).

6.6.2. التحلل الحيوي الفعال للمخلفات البلاستيكية الإلكترونية

ويبلغ الجزء البلاستيكي من المخلفات الإلكترونية حوالي 37%. تشتمل هذه المواد البلاستيكية تقريبًا على 15 نوعًا مختلفًا من البوليمرات ذات التركيبة التالية: 27% بوليسترين، 24% أكريلونيتريل، 7% بولي كربونات، 5% بولي بروبيلين، 2% بولي إيثيلين، ومزيج من اللدائن الحرارية والبوليمرات الأخرى التي تشكل الجزء المتبقي. يمكن إخضاع بعض هذه المواد البلاستيكية لإعادة التدوير؛ ومع ذلك، فإن تحللها البيولوجي يمكن أن يوفر استراتيجية صديقة للبيئة. تم دراسة العديد من الكائنات الحية الدقيقة كمحلل للبلاستيك (Ru et al. 2020; Ali et al. 2021a). أظهرت الكائنات الحية الدقيقة المختلفة التي تم عزلها من مواقع مختلفة بما في ذلك

sites, waste sludge, and marine water showed potent capacity to degrade synthesized polymers such as polyvinyl chloride, polystyrene, and polypropylene to their simple monomeric units in some cases (Raddadi and Fava 2019). The depolymerization of synthetic polymers using microbes is often depending on some microbial enzymes that can target oxidizing the C-C bonds such as esterases, laccases, hydrolases, and peroxidases (Amobonye et al. 2021; Othman et al. 2021).

Plastic biodegradation is naturally a very slow process, however, the improving of the process could be supported by the biological engineering techniques that can play an important role in improving the efficiency of synthetic polymer biodegradation. For instance, the protein engineering technology can be directed to produce modified enzymes with potent catalytic activity towards the degradation of synthetic polymers (Tournier et al. 2020). The using of mature synthetic-biology approaches such as sequencing of whole genome, multi-omics, and obtaining novel strain

مواقع مدافن المخلفات وحمأة المخلفات والمياه البحرية قدرة قوية على تحليل البوليمرات المصنعة مثل كلوريد البوليفينيل والبوليستيرين والبولي بروبيلين إلى وحداتها الأحادية البسيطة في بعض الحالات (Raddadi and Fava 2019). غالبًا ما تعتمد إزالة بلمرة البوليمرات الاصطناعية باستخدام الميكروبات على بعض الإنزيمات الميكروبية التي يمكن أن تستهدف أكسدة روابط CC مثل الاستريزات واللاكيز والهيدروليز والبيروكسيديز (Amobonye et al. 2021; Othman et al. 2021).

يعد التحليل الحيوي للبلاستيك بطبيعة الحال عملية بطيئة للغاية، ومع ذلك، يمكن دعم تحسين العملية من خلال تقنيات الهندسة البيولوجية التي يمكن أن تلعب دورًا مهمًا في تحسين كفاءة التحلل الحيوي للبوليمر الاصطناعي. على سبيل المثال، يمكن توجيه تكنولوجيا هندسة البروتين لإنتاج إنزيمات معدلة ذات نشاط تحفيزي قوي تجاه تحلل البوليمرات الاصطناعية (Tournier et al. 2020). إن استخدام أساليب البيولوجيا التخليقية الناضجة، مثل تسلسل الجينوم الكامل، والأوميات المتعددة، والحصول على بناء سلالة جديدة

من خلال أداة تحرير الجينوم، يمكن أن يساعد الميكروب على اكتساب وظائف جديدة أو تعزيز نشاطه نحو التحلل الحيوي للبوليمرات الاصطناعية (Jaiswal et al. 2020). بالإضافة إلى ذلك، يمكن أن تساعد أدوات المعلوماتية الحيوية أو البيولوجيا الحسابية أيضًا في التحلل الحيوي المعزز للمواد البلاستيكية مثل استخدام التنبؤات القائمة على النماذج لطبيعة الارتباط بين الإنزيم والبوليمر (Skariyachan et al. 2022). لقد تم اقتراح أن هندسة الثقافات الميكروبية المختلطة أو الاتحادات للقيام بمهام ووظائف معقدة أفضل من تصميم وتحسين جميع الجينات المطلوبة في سلالة ميكروبية واحدة أو سلالة ميكروبية واحدة فقط (**الشكل 11**) (McCarty and Ledesma-Amaro 2019). وفقًا لهذا، تم الإبلاغ عن أن الاتحادات الميكروبية المصممة خصيصًا كانت قادرة على التحلل الحيوي لبوليمرات PE وPET بشكل أفضل من السلالات الفردية. نظرًا لأن الجمع بين سلالات ميكروبية متعددة قد يوفر تأثيرًا تآزريًا للتحلل البيولوجي (Jaiswal et al. 2020; Skariyachan et al. 2022).

construction through the genome-editing tool can help the microbe to gain new functions or boost their activity towards synthetic polymers biodegradation (Jaiswal et al. 2020). In addition, the bioinformatics or computational biology tools can also assist for the enhanced biodegradation of plastics such as using the model-based predictions of the nature of binding between the enzyme and the polymer (Skariyachan et al. 2022). It has been proposed that the engineering of mixed microbial cultures or consortia to do complex tasks and functionalities is better than designing and optimizing all required genes in single or only one microbial strain (Figure 11) (McCarty and Ledesma-Amaro 2019). According to this, it has been reported that tailored microbial consortia were able to effectively biodegrade PE and PET polymers better than the individual strains. As the combination among multiple microbial strains may offer synergistic biodegradation effect (Jaiswal et al. 2020; Skariyachan et al. 2022).

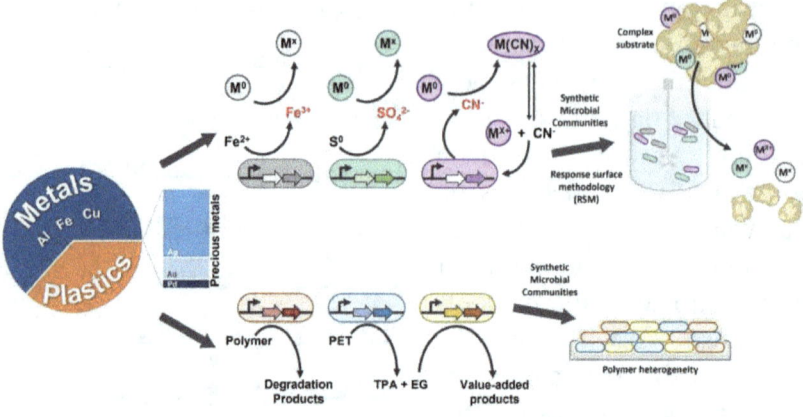

Figure 11 Valorized bioremediation of e-waste using microbes (Han et al. 2022).

الشكل 11 المعالجة الحيوية ذات القيمة للمخلفات الإلكترونية باستخدام الميكروبات (Han et al. 2022).

7. Microbial metabolites used for bioremediation

There are some metabolites that are produced by microbial strains and also help for the bioremediation of pollutants. Biosurfactants, organic acids, and polymeric structures are examples of these metabolites. Biosurfactants such as polymixin, glycoproteins, viscosin, and gramicidin are produced by microbes in order to increase the bioavailability of hydrophobic substances such as poly aromatic hydrocarbons through their solubilization and mobilization (Saha et al. 2021). Moreover, organic

7. المنتجات الايضية الميكروبية المستخدمة في المعالجة الحيوية

هناك بعض المركبات الايضية التي تنتجها السلالات الميكروبية تساعد في المعالجة الحيوية للملوثات. تعتبر المواد الخافضة للتوتر السطحي والأحماض العضوية والهياكل البوليمرية أمثلة على هذه المركبات الايضية. يتم إنتاج المواد الخافضة للتوتر السطحي مثل البوليمكسين والبروتينات السكرية والفيسكوزين والجراميسيدين بواسطة الميكروبات من أجل زيادة التوافر الحيوي للمواد الكارهة للماء مثل الهيدروكربونات العطرية المتعددة من خلال إذابتها وتعبئتها (Saha

acids such as acetic, malic, and citric acids are improving the solubility and mobility of metals and improve their bioremediation (Saha et al. 2021). In addition, polymeric substances such as polysaccharides and polyphosphates are also enhancing the bioremediation process through the phytostabilization and mobility of the metals (Ayilara and Babalola 2023).

8. Production of value-added products from waste streams

In a lot of cases, the wastes are biodegraded and fermented by microbes into value-added bio-products such as organic acids, bio-pigments, proteins, fatty acids, carbohydrates, enzymes, and biofuels (Zabermawi et al. 2022a). Some biomass wastes such as cotton and corn stover have been used by *Streptomyces fulvissimus* as raw materials for the stimulation of the production of some catalytic enzymes such as pectinases, cellulases, and xylanases in order to biodegrade these wastes (Adegboye et al. 2021). The enzymatic activities of laccase, lignin peroxidase, and

(et al. 2021). علاوة على ذلك، تعمل الأحماض العضوية مثل أحماض الخليك والماليك والستريك على تحسين قابلية ذوبان المعادن وحركتها وتحسين معالجتها الحيوية (Saha et al. 2021). بالإضافة إلى ذلك، تعمل المواد البوليمرية مثل السكريات والفوسفات على تعزيز عملية المعالجة الحيوية من خلال التثبيت النباتي وتنقل المعادن (Ayilara and Babalola 2023).

8. إنتاج منتجات ذات قيمة من مجاري المخلفات

في كثير من الاحيان، تتحلل المخلفات بيولوجيًا وتتخمر بواسطة الميكروبات إلى منتجات حيوية ذات قيمة عالية مثل الأحماض العضوية والأصباغ الحيوية والبروتينات والأحماض الدهنية والكربوهيدرات والإنزيمات والوقود الحيوي (Zabermawi et al. 2022a). تم استخدام بعض مخلفات الكتلة الحيوية مثل القطن وحطب الذرة بواسطة *Streptomyces fulvissimus* كمواد خام لتحفيز إنتاج بعض الإنزيمات التحفيزية مثل البكتيناز، والسيلولاز، والزيلاناز من أجل التحليل الحيوي لهذه

manganese peroxidase produced by the white-rot fungi *Pleurotus ostreatus* and *Pleurotus eryngii* have been used to obtain a value-added hydrolysate through the de-polymerization of lignin waste (Melanouri et al. 2022). This hydrolysate which is mainly sugars have been converted into biofuels such as biohydrogen, biomethane, bioethanol, and biobutanol, in addition to organic acids such as acetic, propionic, and lactic acids (Kalayu 2019).

Organic wastes in addition to carbohydrates-rich biomass, are good sources to be bio-remediated into value-added products especially biofuels. Moreover, the plant materials that have high sugar content can be also used as raw materials for the creation of value-added products during the bioremediation process (Table 3). However, in some cases, the inorganic wastes such as heavy metals can be also used by certain microbes in order to create value-added products as subsidiary products. For instance, *Bacillus* sp., has been shown to produce Mn^{2+} oxidizing enzyme that was able to indirectly oxidize the

المخلفات (Adegboye et al. 2021). تم استخدام الأنشطة الأنزيمية للاكيز واللجنين بيروكسيديز وبيروكسيداز المنغنيز التي تنتجها فطريات *Pleurotus ostreatus* و *Pleurotus eryngii* للحصول على هيدروليزات ذات قيمة عالية من خلال إزالة بلمرة مخلفات اللجنين (Melanouri et al. 2022). تم تحويل هذا الهيدروليزات الذي يتكون أساسًا من السكريات إلى وقود حيوي مثل الهيدروجين الحيوي والميثان الحيوي والإيثانول الحيوي والبيوتانول الحيوي، بالإضافة إلى الأحماض العضوية مثل أحماض الأسيتيك والبروبيونيك واللاكتيك (Kalayu 2019).

تعتبر المخلفات العضوية بالإضافة إلى الكتلة الحيوية الغنية بالكربوهيدرات مصادر جيدة للمعالجة الحيوية وتحويلها إلى منتجات ذات قيمة عالية وخاصة الوقود الحيوي. علاوة على ذلك، يمكن أيضًا استخدام المواد النباتية التي تحتوي على نسبة عالية من السكر كمواد خام لإنشاء منتجات ذات قيمة عالية أثناء عملية المعالجة الحيوية (الجدول 3). ومع ذلك، في بعض الحالات، يمكن أيضًا استخدام المخلفات غير العضوية مثل المعادن الثقيلة

chromium III ion into chromium VI ions which are the mobile and bioavailable form of the metal with simultaneous production of sulfuric acid as a value-add by-product (Ghosh et al. 2023).

بواسطة ميكروبات معينة لإنشاء منتجات ذات قيمة عالية كمنتجات فرعية. على سبيل المثال، ثبت أن Bacillus sp تنتج إنزيمًا مؤكسدًا لأيونات Mn^{2+} والذي له القدرة على أكسدة أيون الكروم III بشكل غير مباشر إلى أيونات الكروم VI التي تعد الشكل المتحرك والمتوفر بيولوجيًا للمعدن مع الإنتاج المتزامن لحمض الكبريتيك كقيمة عالية كمنتج ثانوي (Ghosh et al. 2023).

Table 3 Bioremediation of solid wastes into value-added products (Ghosh et al. 2023).

الجدول 3 المعالجة الحيوية للمخلفات الصلبة وتحويلها إلى منتجات ذات قيمة عالية (Ghosh et al. 2023).

Solid waste المخلفات الصلبة	Microorganism الكائن الحي الدقيق	Solid waste and/or produced enzyme المخلفات الصلبة و/أو الإنزيم المنتج	Value-added product منتجات ذات قيمة عالية	References المراجع
Agriculture solid waste (lignocellulose biomass) المخلفات الصلبة الزراعية (الكتلة الحيوية اللجنوسيليلو)	Bacillus pumilus	Xylanase	Ethanol	(Dar et al. 2022)
	Rhizopus oryzae	Lipases	Methanol, ethanol	(López-Fernández et al. 2021)
	Aspergillus oryzae	β-xylanase, alkaline protease, α-amylases, glucoamylases	Ethanol	(Melnichuk et al. 2020)
	Trichoderma viride	Endoglucanase, β-glucosidase, cellulose	Butanol, ethanol, methane, single cell protein	(Anichebe et al. 2019)
Organic waste & food waste (phosphorus rich sewage sludge,	Pleurotus eryngii	Peroxidase, endoglucanase, laccase, lignin peroxidase, manganese xylanases	Flavouring agent (ferulic acid, vanillic acid)	(Melanouri et al. 2022)

animal remains) المخلفات العضوية ومخلفات الطعام (حمأة الصرف الصحي الغنية بالفوسفور وبقايا الحيوانات)			nitrogen reached compost	
	Aspergillus fumigatus	Hemicellulose, Cellulase السليلوز والهيميسليلوز	Ethanol	(Jin et al. 2020)
	Hanseniaspora uvarum (STDF-B2)	Office paper waste مخلفات المكاتب الورقية	Ethanol	(Mansy et al. 2024)
	Aspergillus niger	Proteases	Amino acid	(Angajala et al. 2022)
	Gluconacetobacter xylinus	Algal biomass الكتله الحيوية للطحالب	Biocellulose	(AH Ibrahim et al. 2020)
	Komagataeibacter hansenii AS.5	Chicken feather ريش الدجاج	Biocellulose	(Goda et al. 2022)
	Pseudomonas sp.	Succinic acid, formic acid, gluconic acid, formic acid, butyrate	Calcium and phosphate biofertilize	(Kalayu 2019)
	Clostridium sporogens	Leucine 2,3-aminomutase	Isobutyric acid	(Bardhan et al. 2019)
Plastic solid waste (polyethylene, polymers) المخلفات البلاستيكية الصلبة (البولي إيثيلين والبوليمرات)	*Chryseobacterium luteola*	Cellulase	Biodiesel, ethanol	(Dabbagh et al. 2019)
	Micrococcus luteus	Oxidoreductase, PETase, esterase, cutinase	Glyoxylic acid, carboxylic acid, glycolic acid	(Mohanan et al. 2020)
	Rhodococcus rhodochrous	Oxidoreductase, PETase, esterase, cutinase	Glyoxylic acid, carboxylic acid, glycolic acid	(Mohanan et al. 2020)
Industrial solid waste (petroleum hydrocarbon trash, heavy metal waste) المخلفات الصناعية الصلبة (مخلفات الهيدروكربونات النفطية، مخلفات المعادن الثقيلة)	*Thiobacillus novellus*	Acetic acid, lactic acid, butyric acid,	Sulphuric acid, Polythionic acid	(Ul-Abdin et al. 2022)

9. Energy production via microbial fuel cell (MFC)

9.1. Principles of MFCs

Microbial fuel cell (MFC) is and emerging outstanding technology that is implemented for renewable energy production with simultaneous bioremediation of wastes. It principally depending on the conversion of the chemical energy stored in the wastes into electrical power by the action of microorganism (Zamri et al. 2023). Its configuration is mainly composed of the anodic and cathode chamber that are connected together through an external wire and internal ion exchange membrane. The environmental wastes which are considered the substrates are oxidized by the microbes in the anodic chamber under anaerobic conditions. The electrode at the anodic chamber is receiving the electrons production through the oxidation process and pass them through the electrical wire that contains a resistor to the cathode chamber. These electrons are combined with oxygen molecules from the air

9. إنتاج الطاقة عن طريق خلايا الوقود الميكروبية (MFC)

9.1. مبادئ MFCs

تعتبر خلية الوقود الميكروبية (MFC) إحدى التقنيات الناشئة المتميزة التي يتم تنفيذها لإنتاج الطاقة المتجددة مع المعالجة الحيوية المتزامنة للمخلفات. تعتمد بشكل أساسي على تحويل الطاقة الكيميائية المخزنة في المخلفات إلى طاقة كهربائية بفعل الكائنات الحية الدقيقة (Zamri et al. 2023). يتكون بشكل أساسي من غرفة الأنوديك والكاثود التي يتم توصيلها معًا من خلال سلك خارجي وغشاء التبادل الأيوني الداخلي. تتأكسد المخلفات البيئية بواسطة الميكروبات الموجودة في الغرفة الأنودية تحت الظروف اللاهوائية. يقوم القطب الموجود في الغرفة الأنودية باستقبال إنتاج الإلكترونات من خلال عملية الأكسدة وتمريرها عبر السلك الكهربائي الذي يحتوي على مقاومة إلى غرفة الكاثود. يتم دمج هذه الإلكترونات مع جزيئات الأكسجين من الهواء وبروتونات الهيدروجين التي تمر عبر غشاء التبادل الأيوني وتشكل الماء في غرفة الكاثود

and the hydrogen protons that pass through the ion exchange membrane and forming water at the aerobic cathode chamber (Figure 12) (Ieropoulos and Greenman 2023).

الهوائية (الشكل 12) (Ieropoulos and Greenman 2023).

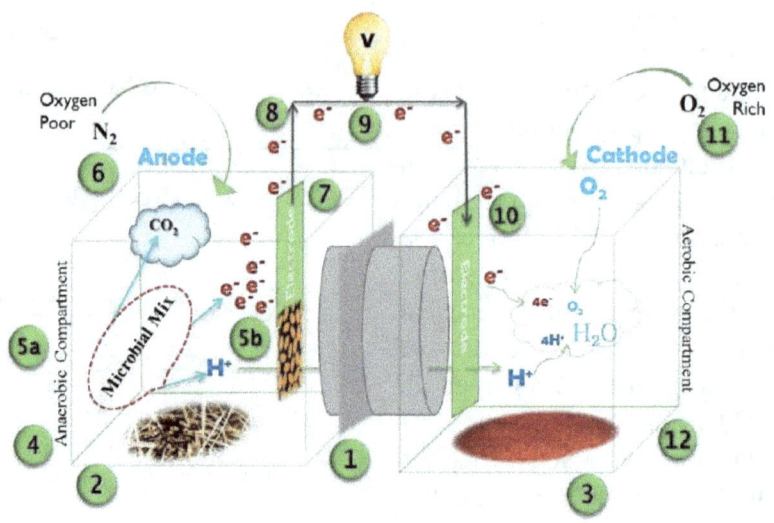

Figure 12 Principles of MFC's working (Zamri et al. 2023).
الشكل 12 مبادئ عمل MFC (Zamri et al. 2023).

9.2. Environmental wastes as substrates for MFC systems

One of the important factors for the efficiency of MFC is the type of the used substrate in addition to its concentration as it significantly impacts the electricity production. Simple and pure components such as glucose and acetate have been used since long time ago as the typical carbon sources. In general, the glucose substrate

9.2. المخلفات البيئية كركائز لأنظمة MFC

أحد العوامل المهمة لكفاءة MFC هو نوع الركيزة المستخدمة بالإضافة إلى تركيزها حيث أنه يؤثر بشكل كبير على إنتاج الكهرباء. تم استخدام مكونات بسيطة ونقية مثل الجلوكوز والأسيتات كمصادر نموذجية للكربون. بشكل عام، يمكن أن تؤدي ركيزة الجلوكوز إلى إنتاج 0.49

can result in the production of 0.49 W/m². It has been also reported that the furan derivatives and the phenolic compounds that are resulted from the hydrolysis of lingo-cellulosic wastes have been used as nutrients in single-chamber MFC. However, the electricity produced by phenolic compounds are much lower than the obtained using glucose as a substrate (Ahanchi et al. 2022).

In other studies, different organic wastes such as industrial wastes, synthetic wastewater, and agricultural biomass have been used as oxidizable substrates for the production of bioelectricity (Fadzli et al. 2021; Naik and Jujjavarappu 2020). Some other scholars proposed that the combined MFC with other technologies are more promising in terms of energy production rather than using standalone MFC systems. For instance, MFC has been combined with constructed wetlands and in other cases with anaerobic membrane bioreactors. This integration is providing the advantages of having aerobic and anaerobic systems that would enhance the outcome benefits including

واط/م2. تم التوصل أيضًا الى أن مشتقات الفوران والمركبات الفينولية الناتجة عن التحلل المائي للنفايات السليلوزية اللغوية قد استخدمت كمواد مغذية في MFC ذات الحجرة الواحدة. ومع ذلك، فإن الكهرباء التي تنتجها المركبات الفينولية أقل بكثير من تلك التي يتم الحصول عليها باستخدام الجلوكوز (Ahanchi et al. 2022).

وفي دراسات أخرى، تم استخدام المخلفات العضوية المختلفة مثل المخلفات الصناعية ومياه الصرف الاصطناعية والكتلة الحيوية الزراعية كركائز قابلة للأكسدة لإنتاج الكهرباء الحيوية (.Fadzli et al 2021; Naik and Jujjavarappu 2020). اقترح بعض العلماء الآخرين أن MFC المدمج مع التقنيات الأخرى يعد واعداً من حيث إنتاج الطاقة بدلاً من استخدام أنظمة MFC المستقلة. على سبيل المثال، تم دمج MFC مع الأراضي الرطبة المشيدة وفي حالات أخرى مع مفاعلات حيوية غشائية لاهوائية. يوفر هذا التكامل مزايا وجود أنظمة هوائية ولاهوائية من شأنها أن تعزز فوائد النتائج بما في ذلك معالجة المخلفات، وإنتاج الغاز

the waste treatment, the biogas production, and the bioelectricity generation (Zhang et al. 2019b).

Recent studies proved the accessibility of food wastes and their resultant leachate as good substrates in MFCs. Both of them has high load of organic matter that can be used as potential resources in MFCs. One of the disadvantages of leachate is its acidic environments that contains a lot of heavy metals and hence can prevent the microbial growth. However, more interested studies reported that the gradual increase of the pH of the anodic chamber to become alkaline has resulted in the gradual increase of the power density to a maximum level (Ahanchi et al. 2022).

Furthermore, food wastes have also been used for the production of activated carbon that can be worked as the anodic electrode instead of platinum wires in order to reduce the total cost of the system. For instance, the spent coffee waste was chemically pretreated with simultaneous carbonization through the pyrolysis process in order to

الحيوي، وتوليد الكهرباء الحيوية (Zhang et al. 2019b).

أثبتت الدراسات الحديثة إمكانية الوصول إلى مخلفات الطعام والعصارة الناتجة عنها كركائز جيدة في الخلايا الجذعية السرطانية. كلاهما يحتوي على كمية كبيرة من المواد العضوية التي يمكن استخدامها كموارد محتملة في الخلايا الجذعية السرطانية. أحد عيوب المادة المرتشحة هي بيئتها الحمضية التي تحتوي على الكثير من المعادن الثقيلة وبالتالي يمكن أن تمنع نمو الميكروبات. ومع ذلك، فقد أفادت دراسات أكثر اهتمامًا أن الزيادة التدريجية في الرقم الهيدروجيني للغرفة الأنودية لتصبح قلوية قد أدت إلى الزيادة التدريجية في كثافة الطاقة إلى الحد الأقصى (Ahanchi et al. 2022).

علاوة على ذلك، تم استخدام مخلفات الطعام أيضًا لإنتاج الكربون النشط الذي يمكن استخدامه كقطب أنوديك بدلاً من أسلاك البلاتين من أجل تقليل التكلفة الإجمالية. على سبيل المثال، تمت معالجة مخلفات القهوة المستهلكة كيميائيًا عن طريق الكربنة المتزامنة من خلال عملية الانحلال الحراري من أجل تحويلها إلى

be transformed into activated carbon that would be worked as the anodic electrode (Figure 13) (Hung et al. 2019).

كربون منشط يمكن استخدامه كقطب أنوديك (الشكل 13) (Hung et al. 2019).

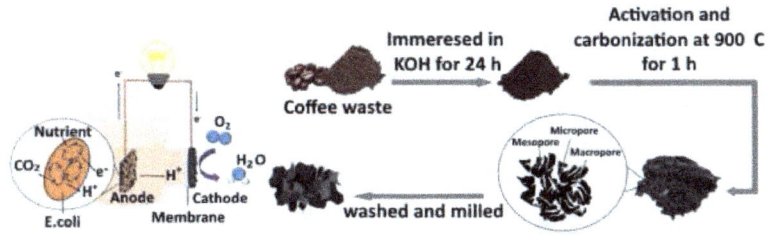

Figure 13 Valorization of spent coffee waste for the preparation of anodic electrodes (Hung et al. 2019).

الشكل 13 تثمين مخلفات القهوة المستهلكة لإعداد الأقطاب الكهربائية الأنودية (Hung et al. 2019).

Some substrates are considered as microbial-rich substrates that leads to promising MFC activity. Cattle manure are one of these substrates. It is rich in both the organic matter and the microbial content which resulted in the benefits of bioremediation of the waste-containing endocrine compounds in addition to the generation of bioelectricity. However, the presence of the methanogenic microbes can inhibit the growth of the electro-active microbes that required for the energy production in MFCs. So, a pretreatment of the cattle manure with heating, chemical

تعتبر بعض الأوساط بمثابة أوساط غنية بالميكروبات تؤدي إلى نشاط MFC. روث الماشية هي واحدة من هذه الأوساط. وهو غني بكل من المادة العضوية والمحتوى الميكروبي مما يؤدى إلى فوائد المعالجة الحيوية للمركبات الصماء المحتوية على المخلفات بالإضافة إلى توليد الطاقة الكهربائية الحيوية. ومع ذلك، فإن وجود الميكروبات الميثانوجينية يمكن أن يمنع نمو الميكروبات النشطة كهربائيًا اللازمة لإنتاج الطاقة في MFC. لذا، فإن المعالجة المسبقة لروث الماشية بالتسخين، أو المعالجة الكيميائية، أو الإشعاع فوق الصوتي ضرورى جدا. أدى استخدام

pretreatment, or ultrasonic radiation is required. Using of at least one of these pretreatments has resulted in production of maximum power density of 12.75 mW/m^2 at 25°C with simultaneous removal of 80% of the COD at the end of the experiment (Syed et al. 2022).

In another study, Ren and his team designed MFC and wetland systems that used swine wastewater as the anodic substrate combined with multi-set air cathodes. At the end of the experiment, they found that these integrated systems resulted in 72% removal of the COD with a power generation of 33.3 mW/m^3 (Ren et al. 2021).

Moreover, MFC integrated moving bed biofilm reactor has included sludges without pre-treatment for 22 days. Approximate amount of 65% of the COD has been removed with simultaneous production of 520 mV bioelectricity. It is believed that the microbial content of the anaerobic bioreactor is stimulating the dehydrogenase activity and resulted in maximized power compared with other systems (Chen et al. 2020). Other

واحدة على الأقل من هذه المعالجات المسبقة إلى إنتاج أقصى كثافة طاقة تبلغ 12.75 ميجاوات/م2 عند 25 درجة مئوية مع إزالة متزامنة لـ 80% من COD في نهاية التجربة (Syed et al. 2022).

وفي دراسة أخرى، صمم Ren وفريقه أنظمة MFC والأراضي الرطبة التي تستخدم مياه الصرف الصحي للخنازير كوسط أنودي مدمج مع كاثودات هوائية متعددة المجموعات. وفي نهاية التجربة، وجدوا أن هذه الأنظمة المتكاملة أدت إلى إزالة 72% من COD مع توليد طاقة قدره 33.3 ميجاوات/م3 (Ren et al. 2021).

علاوة على ذلك، يتضمن مفاعل الأغشية الحيوية ذات الطبقة المتحركة المتكاملة MFC الحمأة دون معالجة مسبقة لمدة 22 يومًا. تمت إزالة الكمية التقريبية البالغة 65% من COD مع الإنتاج المتزامن للكهرباء الحيوية بقدرة 520 مللي فولت. من المعتقد أن المحتوى الميكروبي للمفاعل الحيوي اللاهوائي يحفز نشاط نزع الهيدروجين ويؤدي إلى زيادة الطاقة إلى الحد الأقصى مقارنة بالأنظمة الأخرى

substrates such as seafood wastewater has also been used as MFC substrates (Jamal et al. 2020). However, in terms of power maximization, the power densities of the MFCs can be increased by 12 times when parallel MFC chambers are used. In addition, the output voltage can be also boosted through the using of power management system (PMS). It has been reported that a value of <0.3 V resulted from MFC stack has been increased to 2.5 V via a PMS. However, more studies and intensive research are required for efficient increasing of the produced bioelectricity by this technology with simultaneous considering of the required cost to develop this system (Kim et al. 2019).

(Chen et al. 2020). كما تم استخدام أوساط أخرى مثل مياه الصرف الصحي للمأكولات البحرية كأوساط MFC (Jamal et al. 2020). ومع ذلك، فيما يتعلق بتعظيم الطاقة، يمكن زيادة كثافات الطاقة لخلايا MFC بمقدار 12 مرة عند استخدام غرف MFC المتوازية. بالإضافة إلى ذلك، يمكن أيضًا تعزيز جهد الخرج من خلال استخدام نظام إدارة الطاقة (PMS). تم التوصل الى أن قيمة <0.3 فولت الناتجة عن مكدس MFC قد تمت زيادتها إلى 2.5 فولت عبر نظام PMS. ومع ذلك، هناك حاجة إلى مزيد من الدراسات والأبحاث المكثفة لزيادة كفاءة الطاقة الكهربائية الحيوية المنتجة بواسطة هذه التكنولوجيا مع الأخذ في الاعتبار في الوقت نفسه التكلفة المطلوبة لتطوير هذا النظام (Kim et al. 2019).

10. Latest advances in microbial bioremediation

10.1. Microbial glycoconjugates

These substances are produced by microbes in order to reduce the surface tension and help in increasing the bioavailability of the pollutants in addition to creating a solvent interface of the organic contaminants. All these factors are strongly enhancing the removal of the intended pollutants from the environment (Bhatt et al. 2021b). Microbial glycoconjugates from *Scedosporium* sp. and *Acinetobacter* sp. have been used for the biodegradation of hydrocarbons of the petroleum waste (Figure 14).

10. أحدث التطورات في المعالجة الحيوية الميكروبية

10.1. الجليكوكونجات الميكروبية

يتم إنتاج هذه المواد بواسطة الميكروبات من أجل تقليل التوتر السطحي والمساعدة في زيادة التوافر الحيوي للملوثات بالإضافة إلى خلق واجهة مذيبة للملوثات العضوية. كل هذه العوامل تعزز بقوة إزالة الملوثات المقصودة من البيئة (Bhatt et al. 2021b). جليكوكونجات الميكروبية من *Scedosporium* sp و *Acinetobacter* sp تم استخدامها للتحلل الحيوي للهيدروكربونات من المخلفات البترولية (الشكل 14).

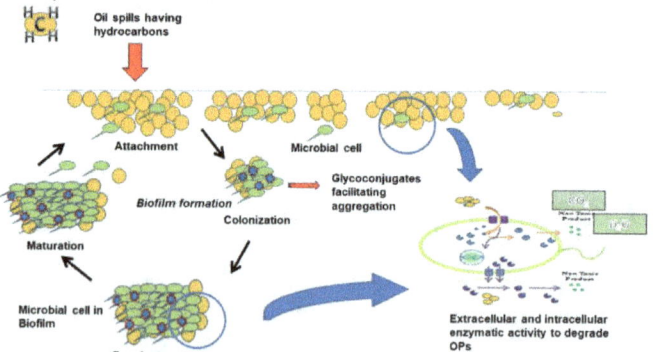

Figure 14 Using of microbial glycoconjugates in the biodegradation of oil spill hydrocarbons (Bhatt et al. 2021b).

الشكل 14 استخدام الجليكوكونجات الميكروبية في التحلل الحيوي للهيدروكربونات المتسربة من النفط (Bhatt et al. 2021b).

10.2. Microbial biofilm

In recent years, microbial biofilm has been extensively used for the bioremediation of organic or recalcitrant pollutants. The microbial biofilm is almost consisting of proteins, polysaccharides, and extracellular DNAs (Sonawane et al. 2022). The technology of biofilm formation is presently improving different environmental factors, enhancing the adhesion to surfaces, and finally improving the quorum sensing (Sonawane et al. 2022). Regarding these advantages, *Exiguobacterium profundum* was significantly reducing the amount of arsenic in synthetic wastewater within two days of incubation (Ayilara and Babalola 2023).

10.3. Bioelectrochemical system

Bioelectrochemical system is an emerging technology where both biological and electrochemical methods are combined in order to remediate the pollutants (Ambaye et al. 2023). This process is hugely used to remediate multiple organic

10.2. الأغشية الحيوية الميكروبية

في السنوات الأخيرة، تم استخدام الأغشية الحيوية الميكروبية على نطاق واسع في المعالجة الحيوية للملوثات العضوية أو المتمردة. يتكون الغشاء الحيوي الميكروبي تقريبًا من البروتينات والسكريات والحمض النووي خارج الخلية (Sonawane et al. 2022). تعمل تقنية تكوين الأغشية الحيوية حاليًا على تحسين العوامل البيئية المختلفة، وتعزيز الالتصاق بالأسطح، وأخيرًا تحسين استشعار النصاب (Sonawane et al. 2022). فيما يتعلق بهذه المزايا، كانت بكتيريا *Exiguobacterium profundum* تقلل بشكل كبير من كمية الزرنيخ في مياه الصرف الصحي الاصطناعية خلال يومين من التحضين (Ayilara and Babalola 2023).

10.3. النظام الكهروكيميائي الحيوي

النظام الكهروكيميائي الحيوي هو تقنية ناشئة حيث يتم الجمع بين الطرق البيولوجية والكهروكيميائية من أجل معالجة الملوثات (Ambaye et al. 2023). تُستخدم هذه العملية بشكل كبير لمعالجة الملوثات العضوية المتعددة مثل

pollutants such as petroleum hydrocarbons depending on the cooperative interactions and the syntrophic of the microbial community that already existed in the contaminated environment (Ambaye et al. 2023). Different microbes such as *Thauera* sp., *Rhodococcus* sp., *Pseudomonas* sp., and *Ralstonia* sp. have been involved in the bioremediation of phenanthrene from hydrocarbons polluted sites (Sharma et al. 2020).

10.4. Nanoparticles

Using of nanotechnology in the bioremediation of contaminants is a globally an emerging technology. The synthesis of nanoparticles can be achieved using physical, chemical, and biological sources (Shanmuganathan et al. 2019). A lot of factors are affecting the efficiency of using nanoparticles in bioremediation such as the shape of the obtained nanoparticles, their size, the type of surface coating, and their chemical structure. In addition, some other factors such as the pollutant nature, the media pH and temperature have been also reported. In recent studies, carbon dots nanoparticles gained huge attention in the remediation of different environmental pollutants as they abundant, have low-toxicity, and have unique-optical characteristics (Long et al. 2021).

11. Challenges, future perspectives, and recommendations

In order to reduce the amount of food waste thrown by households and restaurants, the researchers are thinking about the supporting of the new technologies such as the designing of refrigerators that can alert the users about the close expiration date of specific products. In some other cases, there would be SMS or emails about the shopping lists, or suggested recipes, that would help for the waste reduction.

However, according to the daily release of food wastes into the environment, there would be some scenario that would help for the reduction of such wastes. Some strategies can be followed in order to encourage the people to reduce the food waste release. For instance, scientific programs or social media advertising

الحموضة ودرجة الحرارة. في الدراسات الحديثة، اكتسبت الجسيمات النانوية ذات النقاط الكربونية اهتمامًا كبيرًا في معالجة الملوثات البيئية المختلفة لأنها وفيرة، وذات سمية منخفضة، ولها خصائص بصرية فريدة (Long et al. 2021).

11. التحديات والآفاق المستقبلية والتوصيات

من أجل تقليل كمية مخلفات الطعام التي تتخلص منها المنازل والمطاعم، يفكر الباحثون في دعم التقنيات الجديدة مثل تصميم الثلاجات التي يمكنها تنبيه المستخدمين حول تاريخ انتهاء الصلاحية لمنتجات معينة. وفي بعض الحالات الأخرى، قد تكون هناك رسائل نصية قصيرة أو رسائل بريد إلكتروني حول قوائم التسوق، أو وصفات مقترحة، من شأنها أن تساعد في تقليل المخلفات.

ومع ذلك، وفقًا للإطلاق اليومي للمخلفات الغذائية في البيئة، سيكون هناك بعض السيناريوهات التي من شأنها أن تساعد في تقليل هذه المخلفات. يمكن اتباع بعض الحيل لتشجيع الناس على تقليل إطلاق مخلفات الطعام. على سبيل المثال، البرامج العلمية أو إعلانات وسائل التواصل

responsible for teaching the people changing their waste release behavior and raising their awareness about the management of food waste release. In some other cases, there might be some kind of incentive and intrinsic motivation.

Some researchers are recommending the using of microbial enzymes for the biodegradation of environmental wastes; however, in such case, the applied enzymatic degradation process should be resulted in secured and more sustainable environment.

Looking for new novel microbes for the bioremediation of environmental wastes, especially industrial wastes is recommended as they might have potent bioremediation ability better than the already existed ones. In same context, the using of microbial consortia with multiple specifies for different substrates could consequently increasing the bioremediation rate and remove multiple pollutants at the same time. In addition, the currently existed microbes can be genetically modified in order to enhance

الاجتماعي المسؤولة عن تعليم الناس تغيير سلوكهم في إطلاق المخلفات وزيادة وعيهم حول إدارة إطلاق المخلفات الغذائية. وفي بعض الحالات الأخرى، قد يكون هناك نوع من الحوافز والدوافع الجوهرية.

ويوصي بعض الباحثين باستخدام الإنزيمات الميكروبية للتحلل الحيوي للمخلفات البيئية؛ ومع ذلك، في مثل هذه الحالة، ينبغي أن تؤدي عملية التحلل الأنزيمي المستخدمه إلى بيئة آمنة وأكثر استدامة.

ويوصى أيضاً بالبحث عن ميكروبات جديدة للمعالجة الحيوية للمخلفات البيئية، وخاصة المخلفات الصناعية لأنها قد تتمتع بقدرة قوية على المعالجة الحيوية أفضل من تلك الموجودة بالفعل. وفي نفس السياق، فإن استخدام اتحادات ميكروبية ذات مواصفات متعددة لركائز مختلفة يمكن أن يؤدي بالتالي إلى زيادة معدل المعالجة الحيوية وإزالة الملوثات المتعددة في نفس الوقت. بالإضافة إلى ذلك، يمكن تعديل الميكروبات الموجودة حاليًا وراثيًا من أجل

their enzymatic production that would result in potent bioremediation process.

It is also recommended to discover the microbes that able to degrade the inorganic or synthetic pollutants with simultaneous degradation of the organic ones, as most of the currently existed microbes are degrading the organic load and leaving out the inorganic load without treatment.

In some other cases, the bioremediation-used microbes are source of contamination instead of removing the existed pollutants, as which happened with the microbial biostimulation strategy that resulting in the algal blooms. In such case, new methods that would help for preventing this phenomenon is required in order to keep the sustainability of the environment.

Investigation of the potency of the biodegradation of pollutants by microbes in the field with its naturally measured temperature is also needed, as most of microbes can work perfectly in laboratories with optimized temperatures, but failed to

تعزيز إنتاجها الأنزيمي مما قد يؤدي إلى عملية معالجة حيوية قوية.

يوصى أيضًا باكتشاف الميكروبات القادرة على تحليل الملوثات غير العضوية أو الاصطناعية مع التحلل المتزامن للملوثات العضوية، حيث أن معظم الميكروبات الموجودة حاليًا تعمل على تحليل الحمل العضوي وترك الحمل غير العضوي دون معالجة.

وفي بعض الحالات الأخرى تكون الميكروبات المستخدمة في المعالجة الحيوية مصدراً للتلوث بدلاً من إزالة الملوثات الموجودة، كما حدث مع استراتيجية التحفيز الحيوي الميكروبي التي أدت إلى ازدهار الطحالب. وفي مثل هذه الحالة لا بد من إيجاد أساليب جديدة تساعد في الوقاية من هذه الظاهرة من أجل الحفاظ على استدامة البيئة.

هناك حاجة أيضًا إلى دراسة فعالية التحلل الحيوي للملوثات بواسطة الميكروبات في الحقل مع درجة الحرارة المقاسة طبيعيًا، حيث يمكن لمعظم الميكروبات العمل بشكل مثالي في المختبرات ذات درجات الحرارة

work similarly under the field conditions.

The follow up of the microbes introduced into the environment in a specific bioremediation mission is also essential, as some of these microbes might become means of pollution if they mutated or genetically modified for instance.

Some microbes, or their enzymes, might die or lose their activities before the starting of the remediation process, so, immobilization of the microbes or the enzymes inside polymeric capsules or even the combination of the enzymes with nanoparticles might protect them from the harsh environmental conditions that might affect their activity. On the other hand, in order to apply the MFC in large scale treatment of wastewater and bioelectricity production, some parameter such as the external resistance capacity and the hydraulic retention time should be optimized, as they are not considered through the laboratory work.

المثلى، لكنها فشلت في العمل بشكل مماثل في ظل الظروف الميدانية.

كما تعد متابعة الميكروبات التي تدخل إلى البيئة في مهمة معالجة بيولوجية معينة أمراً ضرورياً أيضاً، حيث أن بعض هذه الميكروبات قد تصبح وسيلة للتلوث إذا تحورت أو عدلت وراثياً على سبيل المثال.

قد تموت بعض الميكروبات أو إنزيماتها أو تفقد نشاطها قبل البدء في عملية المعالجة، لذا فإن تثبيت الميكروبات أو الإنزيمات داخل الكبسولات البوليمرية أو حتى دمج الإنزيمات مع الجسيمات النانوية قد يحميها من الظروف البيئية القاسية التي قد يؤثر على نشاطهم من ناحية أخرى، من أجل تطبيق MFC في معالجة واسعة النطاق لمياه الصرف الصحي وإنتاج الكهرباء الحيوية، يجب تحسين بعض المعلمات مثل قدرة المقاومة الخارجية ووقت الاحتفاظ الهيدروليكي، حيث لا يتم أخذها في الاعتبار من خلال العمل المختبري.

12. Egyptian experiences and participations in the field of waste's bioremediation

The environmental bioremediation field has been studied by many Egyptian researchers. Some of them has focused on the reduction of the toxicity of the environmental pollutants into a level that would be safe for the human and other creatures. However, some other researchers were interested in not only the reduction of the amount of the released wastes, but also in maximizing the benefits of these wastes through the obtaining of value-added products.

Herein, some of the research articles that reflect the participation of the Egyptian researchers alone and in participation with researchers from other countries (Figure 15) in the bioremediation of environmental wastes into less toxic or into beneficial products.

Darwesh and his team succeeded to isolate *Fusarium oxysporum* fungus from contaminated sites that was able to synthesize iron

.12 خبرة ومشاركة المصرين في مجال المعالجة الحيوية للمخلفات

تمت دراسة مجال المعالجة الحيوية البيئية من قبل العديد من الباحثين المصريين. وقد ركز بعضها على تقليل سمية الملوثات البيئية إلى مستوى آمن للإنسان والمخلوقات الأخرى. ومع ذلك، اهتم بعض الباحثين الآخرين ليس فقط بتقليل كمية النفايات المنطلقة، ولكن أيضًا بتعظيم فوائد هذه النفايات من خلال الحصول على منتجات ذات قيمة عالية.

وهنا بعض المقالات البحثية التي تعكس مشاركة الباحثين المصريين منفردين وبالمشاركة مع باحثين من دول أخرى (شكل 15) في المعالجة الحيوية للمخلفات البيئية وتحويلها إلى منتجات أقل سمية أو مفيدة، فعلى سبيل المثال:

نجح درويش وفريقه في عزل فطر *Fusarium oxysporum* من المواقع الملوثة، وتمكن من تصنيع جزيئات الحديد النانوية بأحجام تتراوح بين 0.7 إلى 3

nanoparticles with size range of 0.7 to 3 nm. These nanoparticles have been used effectively as antimicrobial agents against multiple environmental pathogens in addition to its ability to remove heavy metals from contaminated wastewater (Darwesh et al. 2021).

Elzakey and his team have also isolated potent microbial species from wastewater samples collected from Al-Khairy agricultural drainage that is receiving both agricultural and domestic wastes. They succeeded to isolate 3 bacterial and 9 fungal species that have effectively bioremediate the residues of chlorpyrifos pesticide with removal percentages 24.16-80.93% at lab scale. They found that the most potent bacterial isolates that can proficiently degrade chlorpyrifos were *Bacillus cereus* PC2 and *Streptomyces praecox* SP1 strains (Elzakey et al. 2023).

Abdel-Razik and his colleagues reported the ability to isolate the halophilic bacteria named *Halomonas* sp. from Lake Qarun in Egypt. The isolate showed high

نانومتر. وقد تم استخدام هذه الجسيمات النانوية بشكل فعال كعوامل مضادة للميكروبات ضد مسببات الأمراض البيئية المتعددة بالإضافة إلى قدرتها على إزالة المعادن الثقيلة من مياه الصرف الصحي الملوثة (Darwesh et al. 2021).

قام الزاكي وفريقه أيضًا بعزل الأنواع الميكروبية القوية من عينات مياه الصرف الصحي التي تم جمعها من مصرف الخيري الزراعي الذي يستقبل المخلفات الزراعية والمنزلية. ونجحوا في عزل 3 أنواع بكتيرية و9 أنواع فطرية قامت بالمعالجة الحيوية لبقايا مبيد الكلوربيريفوس بفعالية وبنسب إزالة 24.16-80.93% على نطاق المختبر. ووجدوا أن أقوى العزلات البكتيرية التي يمكنها تحليل الكلوربيريفوس بكفاءة هي سلالات *Bacillus cereus* PC2 و *Streptomyces praecox* SP1 (Elzakey et al. 2023).

أفاد عبد الرازق وزملاؤه عن القدرة على عزل البكتيريا المحبة للملوحة والتي تسمى *Halomonas* sp. من بحيرة قارون في مصر. أظهرت العزلة مقاومة عالية لـ 4

resistant to 4 mM of Pb in addition to its resistance to other metals. The bacterial isolate succeeded to uptake 94% of the Pb metal during the first 6 h of incubation. The TEM investigation showed that the Pb nanoparticles have been synthesized inside the cells and were accumulated in the area rich with exopolysaccharides (Abdel-Razik et al. 2020).

Zabermawy and his team members isolated two promising bacteria isolates namely *Enterobacter cloacae* 279-56 (R4) and *Pseudomonas otitis* MCC10330 (R19) that were able to effectively biodegrade oil residues from oily industrial wastewater. Bothe isolates were able to degrade the oil content with simultaneous removal of the organic load. They have removed 84 mg/L of the existed oil content which was not enough for the water quality to be within the safe discharging limits. For that reason, and in order to enhance the bioremediation process of oil pollutant, the authors carried out robust experiments with mixed efficient bacterial strains (Zabermawi et al. 2022b).

Gaber and his colleagues have also reported the biodegradation of diazinon pesticide by the wastewater isolated fungal species *Penciillium Citreonigum*, *Aspergillus fumigatus*, and *Rhizopus nodosus*. The biodegradation percentage of diazinon was from 72.2 to 91.1% according to the type of the isolate. The three isolates succeeded to minimize the time required for the biodegradation of the pesticide to be in the range from 7.7 to 12.1 days compared with 87.8 days of the control sample, with the release of 2-Isopropyl-4-methyl-6-hydroxypyrimidine (IMHP) as the main biodegradation byproduct (Gaber et al. 2020).

Radwan and his team succeeded to use the chemically and physically treated *Chlorella vulgaris* for the bioremediation of reactive yellow 145 dye as well as the biosorption of Cu ions. The algal strain has been treated by citric acid that helps for the insertion of carbonyl group into the algal surface, in addition to the intensification

وقد درس جابر وزملاؤه أيضًا التحلل الحيوي لمبيد الديازينون بواسطة الأنواع الفطرية المعزولة من مياه الصرف الصحي *Penciillium Citreonigum*، و*Aspergillus fumigatus*، و*Rizopus nodosus*. وكانت نسبة التحلل الحيوي للديازينون من 72.2 إلى 91.1% حسب نوع العزلة. نجحت العزلات الثلاث في تقليل الوقت اللازم للتحلل الحيوي للمبيد ليكون في حدود 7.7 إلى 12.1 يوم مقارنة بـ 87.8 يوم لعينة السيطرة، مع إطلاق 2-أيزوبروبيل-4-ميثيل-6-هيدروكسي بيريميدين. (IMHP) باعتباره المنتج الثانوي الرئيسي للتحلل الحيوي (Gaber et al. 2020).

نجح رضوان وفريقه في استخدام *Chlorella vulgaris* المعالج كيميائيًا وفيزيائيًا للمعالجة الحيوية للصبغة الصفراء التفاعلية 145 بالإضافة إلى الامتصاص الحيوي لأيونات النحاس. تمت معالجة سلالة الطحالب بحامض الستريك الذي يساعد على إدخال مجموعة الكاربونيل إلى سطح الطحالب، بالإضافة

of the surface functional groups of the cells by the heat treatment. The citric acid modified cells were able to remove 97% of the dye in 40 min. While the heat treated algal cells were the best for the biosorption of 92% of the copper ions in 5 min (Radwan et al. 2020).

Taha and his colleagues succeeded to design a pressure-free filtration system that includes PVA/CS polymeric membrane-integrated silver nanoparticles for the removal of microbes from contaminated water. They have greenly synthesized AgNPs using the filtrate of *Bacillus endophyticus* bacterial culture. They tested that polymeric/nanoparticles composite against six pathogenic microbes. It showed potent activity to kill and prevent the passage of 88% of the original microbial content in the tested samples compared with control (Taha et al. 2019).

Abdel-Razek and his research team have effectively used three bacterial isolates namely *Bacillus* sp., *Bacillus thuringiensis*, and *Enterobacter cloacae* for the bioremediation of phenanthrene from contaminated samples. The GC-MS analysis showed that the formed byproducts of the phenanthrene biodegradation process contained lower carbon atom numbers ranged from C4 to C12, with no cytotoxic effects against Vero cell lines compared with the phenanthrene hydrocarbon itself (Abdel-Razek et al. 2020).

Taha and others has accelerated the treatment of the anthracene-contaminated water through the designing of multiple columns with successive stages of anthracene-graphite adsorption followed by a biodegradation separate step. The have firstly adsorbs the anthracene from the contaminated water on free and immobilized graphite particles in order to have purified water in a short time with removal efficiency of 83.5%. While the absorbed anthracene integrated with the graphite powder were transferred into a new tank that have the microbial strain namely *Bacillus* sp., in order to be degraded into less or non-harmful compounds. The

cloacae للمعالجة الحيوية للفينانثرين من العينات الملوثة. أظهر تحليل GC-MS أن المنتجات الثانوية المتكونة من عملية التحلل الحيوي للفينانثرين تحتوي على أعداد ذرات كربون أقل تتراوح من C4 إلى C12، مع عدم وجود آثار سامة للخلايا ضد خطوط خلايا فيرو مقارنة بهيدروكربون الفينانثرين نفسه (-Abdel Razek et al. 2020).

قام طه وآخرون بتسريع معالجة المياه الملوثة بالأنثراسين من خلال تصميم أعمدة متعددة بمراحل متتالية من امتصاص الجرافيت والأنثراسين تليها خطوة منفصلة للتحلل الحيوي. قامت أولاً بامتصاص الأنثراسين من المياه الملوثة على جزيئات الجرافيت الحرة والثابتة من أجل الحصول على مياه نقية في وقت قصير مع كفاءة إزالة تبلغ 83.5%. بينما تم نقل الأنثراسين الممتص المدمج مع مسحوق الجرافيت إلى خزان جديد يحتوي على السلالة الميكروبية *Bacillus* sp.، وذلك لتحلله إلى مركبات أقل أو غير ضارة. وكانت نسب التحلل الحيوي لكل من مساحيق الأنثراسين الحرة والمجمدة 62.7 و82.6% على التوالي (Tahaa et al. 2021).

biodegradation percentages of both freely and immobilized anthracene powders were 62.7 and 82.6%, respectively (Tahaa et al. 2021).

Abu-Saied with other researchers has been succeeded to synthesize silver nanoparticles in a green way using an extract of the pupa of the green bottle fly insect. The biosynthesized silver nanoparticles were integrated into k-carrageenan polymer in order to be used for the decontamination of microbes from contaminated water. They reported highly effective antimicrobial activity of the polymer-integrated nanoparticles against Gram negative, Gram positive, and yeast strains (Abu-Saied et al. 2020).

Moustafa and his team have tested dried powders of the fruits of *Syzygium aromaticum* (clove) in addition to the leaves and fruits of *Schinus molle* to remediate some heavy metals in addition to ceasing the growth of the microbial pathogens in wastewater. They have found that clove powder was moderately able to remediate nickel, zinc, and cobalt with removal

وقد نجح أبو سعيد مع باحثين آخرين في تصنيع جزيئات الفضة النانوية بطريقة خضراء باستخدام مستخلص يرقات حشرة ذبابة الزجاجة. تم دمج جزيئات الفضة النانوية المُصنَّعة حيويًا في بوليمر الكاراجينان k لاستخدامها في إزالة التلوث من الميكروبات من المياه الملوثة. وقد أبلغوا عن نشاط مضاد للميكروبات فعال للغاية للجسيمات النانوية المدمجة بالبوليمر ضد سلالات سلبية الجرام وإيجابية الجرام والخميرة (Abu-Saied et al. 2020).

قام مصطفى وفريقه باختبار المساحيق المجففة من ثمار نبات *Syzygium Aromaticum* (القرنفل) بالإضافة إلى أوراق وثمار نبات *Schinus molle* لمعالجة بعض المعادن الثقيلة بالإضافة إلى وقف نمو مسببات الأمراض الميكروبية في مياه الصرف الصحي. لقد وجدوا أن مسحوق القرنفل كان قادرًا بشكل معتدل على معالجة النيكل والزنك والكوبالت بنسب إزالة تراوحت من 26.19 إلى 39.8%. بينما أظهرت أوراق نبات

percentages ranged from 26.19 to 39.8%. While the Schinus leaves showed a remediation percentage of 90% of lead ions, that has been elevated to 96.6% when the leaves powder was immobilized inside alginate beads. However, both plants showed remarkable antimicrobial activities against the tested microbial pathogens *Vibrio cholerae*, *Candida albicans*, *Pseudomonas aeruginosa*, *Escherichia coli*, *Klebsiella pneumoniae*, *Staphylococcus aureus* and *Bacillus cereus* (Moustafa et al. 2020).

Moustafa and his team have also been using silver nanoparticles that were greenly prepared by the larvae of the green bottle fly insect for the bioremediation of Imidacloprid pesticide. They have prepared chitosan polymer integrated with silver nanoparticles in the size of 25-50 nm. They found that the chitosan itself can remove approximately 40% of the pesticide, which has been increased to 85% after the integration of the greenly prepared nanoparticles at slightly acidic pH (Moustafa et al. 2021).

الشينوس نسبة معالجة بلغت 90% من أيونات الرصاص، وقد ارتفعت هذه النسبة إلى 96.6% عندما تم تثبيت مسحوق الأوراق داخل حبات الجينات. ومع ذلك، أظهر كلا النباتين أنشطة ملحوظة مضادة للميكروبات ضد مسببات الأمراض الميكروبية في الخصية *Vibrio cholerae, Candida albicans, Pseudomonas aeruginosa, Escherichia coli, Klebsiella pneumoniae, Staphylococcus aureus* and *Bacillus cereus* (Moustafa et al. 2020).

كما استخدم مصطفى وفريقه جزيئات الفضة النانوية التي تم تحضيرها باللون الأخضر بواسطة يرقات حشرة ذبابة الزجاجة للمعالجة الحيوية لمبيد إيميداكلوبريد. لقد قاموا بتحضير بوليمر الكيتوزان المتكامل مع جسيمات الفضة النانوية بحجم 25-50 نانومتر. ووجدوا أن الكيتوزان نفسه يمكنه إزالة ما يقرب من 40% من المبيد، والتي تمت زيادتها إلى 85% بعد دمج الجسيمات النانوية المحضرة باللون الأخضر عند درجة حموضة حمضية قليلاً (Moustafa et al. 2021).

Mostafa and his colleagues have valorized the using of potato wastewater, as low-cost media, for the microbial synthesis of β-cyclodextrin glycosyltransferase with thermal stability trait. The have reported that the encapsulation of the of the enzyme-producing bacteria *Bacillus licheniformis* inside alginate capsules can maximize the enzyme production twice-fold than using the free cells. They have also reported that the starch molecules existed inside the potato wastewater can induce the production of the enzyme with a concentration of 25 U/g of raw potato starch (Mostafa et al. 2021).

Taha and his team members have used the office paper waste as a raw material for the production of bioethanol as an alternative energy form. They have been reported that the optimum physico-chemical and enzymatic hydrolysis can result in the liberation of 7 mg/ml of glucose units after 48 h. These sugars have been subsequently fermented by the yeast *Saccharomyces cerevisiae* into 0.12% bioethanol. They have also reported that the concentration

قام مصطفى وزملاؤه بتقييم استخدام مياه الصرف الصحي للبطاطس، كوسيلة منخفضة التكلفة، للتخليق الميكروبي لـ β-cyclodextrin glycosyltransferase مع سمة الاستقرار الحراري. لقد أفادوا أن تغليف البكتيريا المنتجة للإنزيم *Bacillus licheniformis* داخل كبسولات الجينات يمكن أن يزيد إنتاج الإنزيم إلى الحد الأقصى مرتين مقارنة باستخدام الخلايا الحرة. وقد أفادوا أيضًا أن جزيئات النشا الموجودة داخل مياه الصرف الصحي للبطاطس يمكن أن تحفز إنتاج الإنزيم بتركيز 25 وحدة / جرام من نشا البطاطس الخام (Moustafa et al. 2021).

استخدم طه وأعضاء فريقه نفايات الورق المكتبية كمادة خام لإنتاج الإيثانول الحيوي كشكل من أشكال الطاقة البديلة. لقد تم الإبلاغ عن أن التحلل المائي الفيزيائي والكيميائي الأمثل يمكن أن يؤدي إلى تحرير 7 ملغم / مل من وحدات الجلوكوز بعد 48 ساعة. تم تخمير هذه السكريات لاحقًا بواسطة خميرة *Saccharomyces cerevisiae* إلى 0.12% من الإيثانول الحيوي. وقد أفادوا أيضًا أن تركيز ثاني أكسيد الكربون المنبعث من خلال عملية

of the CO_2 released through the fermentation process can indirectly reflect the exact concentration of the biosynthesized ethanol (Taha et al. 2021). They have also used the cardboard wastes as raw materials for the production of bioethanol. In that case, they have fermented the sugars obtained from the hydrolysis of cardboard waste into 2.9 mg/ml bioethanol using *Pichia nakasei* yeast isolate. They have also succeeded to purify the produced bioethanol from 0.293 to 0.40% using an Amicon cell containing poly(acrylonitrile-co-methyl acrylate) modified with EDA membrane, through the application of nitrogen gas pressure at 40 psi (Taha et al. 2022).

El-Badry and others have used the PCBs, an electronic waste, for the bioleaching of copper metal using *Actinomycete graminofaciens*. The microbial strain has succeeded to leach 88.1% of the copper at pH 5. They have reported that the higher dissolution rate has occurred in the non-heated sample than the heated one. The SEM micrographs revealed the presence of more

التخمير يمكن أن يعكس بشكل غير مباشر التركيز الدقيق للإيثانول المُصنّع حيويًا (Taha et al. 2022).

كما استخدموا نفايات الورق المقوى كمواد خام لإنتاج الإيثانول الحيوي. في هذه الحالة، قاموا بتخمير السكريات التي تم الحصول عليها من التحلل المائي لنفايات الورق المقوى إلى 2.9 ملجم/مل من الإيثانول الحيوي باستخدام خميرة *Pichia nakasei* المعزولة. لقد نجحوا أيضًا في تنقية الإيثانول الحيوي المنتج من 0.293 إلى 0.40% باستخدام خلية أميكون تحتوي على بولي (أكريلونيتريل-ميثيل أكريلات) المعدل بغشاء EDA، من خلال تطبيق ضغط غاز النيتروجين عند 40 رطل لكل بوصة مربعة (Taha et al. 2022).

استخدم البدري وآخرون مركبات ثنائي الفينيل متعدد الكلور، وهي نفايات إلكترونية، في الترشيح الحيوي لمعدن النحاس باستخدام *Actinomycete graminofaciens*. نجحت السلالة الميكروبية في ترشيح 88.1% من النحاس عند درجة حموضة 5. وقد أفادوا أن معدل الذوبان الأعلى قد حدث في العينة غير المسخنة مقارنة بالعينة المسخنة. كشفت

pores in the e-waste that have been formed after the bio-treatment process, which indicating the liberation of the metal into the surrounding solution (El-Badry et al. 2021).

Sharada and his team have also been bioleaching the copper from PCBs using *Aspergillus niger*. They succeeded to extract 100% of copper under optimized conditions. They have been reported that both malic and citric acids with high concentrations were detected in the sample with e-waste than the sample lacking this waste. They have also confirmed the bioleaching process through the appearance of smooth and porous surfaces according to the SEM micrographs (Sharada et al. 2021).

Abdelaal and other researchers reported the phytoremediation of some heavy metals by aquatic macrophytes. They have reported the ability of *Phragmites australis* to accumulate effectively Co, Cd, Fe, and Ni at higher concentrations. They have also reported the ability of *Eichhornia crassipes* to accumulate higher concentrations of Zn, Cu, Pb, and Mn (Abdelaal et al. 2021). In the same context, Elshamy and his colleagues has also reported the phytoremediation of heavy metals by plants. They have used *Portulaca oleracea* to accumulate Pb, Zn, Fe, Cu, and Mn from industrial effluents (Elshamy et al. 2019).

Saleh and his research team have used the starchy components of the kitchen waste as cheap and cost-effective raw materials for the production of bacterial cellulose as a value-added product using *Komagataeibacter hansenii* AS.5 strain. They have used the produced biocellulose (BC) as supporting material for graphite and charcoal in order to be used for the removal of cationic dyes such as methylene blue (MB) from wastewater. Their comparable study revealed the ability of graphite loaded BC and charcoal loaded BC to remove 98.7 and 100% of MB from water compared with 53% removal by the BC alone (Saleh et al. 2021).

قدرة *Eichhornia crassipes* على تجميع تركيزات أعلى من الزنك والنحاس والرصاص والمنغنيز (Abdelaal et al. 2021). وفي السياق نفسه، أفاد الشامي وزملاؤه أيضًا عن المعالجة النباتية للمعادن الثقيلة بواسطة النباتات. وقد استخدموا *Portulaca oleracea* لتجميع الرصاص والزنك والحديد والنحاس والمنغنيز من المخلفات السائلة الصناعية (Elshamy et al. 2019). استخدم صالح وفريقه البحثي المكونات النشوية في نفايات المطبخ كمواد خام رخيصة وفعالة من حيث التكلفة لإنتاج السليلوز البكتيري كمنتج ذي قيمة مضافة باستخدام سلالة *Komagataeibacter hansenii* AS.5. لقد استخدموا السليلوز الحيوي المنتج (BC) كمواد داعمة للجرافيت والفحم لاستخدامه في إزالة الأصباغ الكاتيونية مثل أزرق الميثيلين (MB) من مياه الصرف الصحي. كشفت دراستهم المقارنة عن قدرة الجرافيت المحمل على BC والفحم المحمل على BC على إزالة 98.7 و100% من أزرق الميثيلين من الماء مقارنة بإزالة 53% بواسطة الـ BC وحده (Saleh et al. 2021).

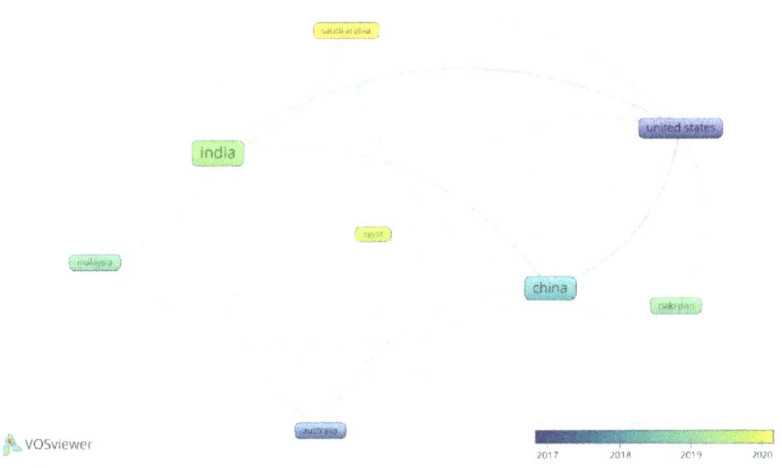

Figure 15 VOSviewer representing the network of Egyptian researchers with other researchers worldwide in the field of bioremediation.

الشكل 15 يمثل شبكة الباحثين المصريين مع باحثين آخرين على مستوى العالم في مجال المعالجة الحيوية بإستخدام برنامج VOSviewer.

13. The Role of Kingdom of Saudi Arabia in Bioremediation Technology

Environmental Biotechnology has been applied to solve wide range of environmental issues such as, pollution using technologies like, Bioremediation. In 1990 AD, the Gluf war caused a huge pollution of Arabian Gulf region with petroleum. After that, the idea of Bioremediation has started to discussed in Saudi Arabia as a solution of oil pollution (Jones et al., 2008, Hashem, 2007).

13. دور المملكة العربية السعودية في تكنولوجيا المعالجة البيولوجية

تم تطبيق التكنولوجيا الحيوية البيئية لحل مجموعة واسعة من القضايا البيئية، مثل قضايا التلوث وذلك باستخدام تقنيات كالمعالجة الحيوية. في عام 1990م، تسببت حرب الخليج بتلوث كبير في منطقة الخليج العربي بالنفط. بعد ذلك، بدأت فكرة المعالجة الحيوية تُناقش في المملكة العربية السعودية كحل للتلوث النفطي (Jones et al., 2008, Hashem, 2007).

Currently, Saudi Arabia has stablished many bioremediation projects and funding them with millions of Saudi Riyals (SR) especially, with reuse the wastewater. For example, one of these projects was done in 2009 at Riyadh (Wadi Hanifah) with budget $32,000,000 and that project was designed by Canadian company (Moriyama & Teshima Planners Ltd). This project was treated an average 92.5 million gallons of urban wastewater daily and this number will reach up to 317 million gallons by 2025 (Series, 2015). A study was done by Saudi group found that *Pseudomonas aeruginosa* strain (AM-1) isolated from Ha'il region has a potential to be applied in biological treatment (Abdulmohsen et al., 2023). Another study by (Moustafa, 2016), found both *Aspergillus niger* and *Lichtheimia ramose* fungi have efficient impact in reducing oil spills from polluted environment. A study by (Al-Dhabi et al., 2020) determined bioremediation of crude oil using *Bacillus subtilis* strain Al-Dhabi-130 that was isolated from the marine soil sediment. The findings showed that *B. subtilis* was able to remove 89% of crude oil.

14. The role of some other Arab countries in bioremediation technology

Middle east and north Africa countries are one of the poorest countries in terms of water availability. Their nature is a harsh desert. The main source of water supply is ground water. One of the main challenges in this region is waster scarcity. This is reflected by the increasing demand in both demotic use as well as agriculture. Agriculture use itself acquires 70% of available water due to the inadequate irrigation techniques. Climate change in the form of higher temperatures and desertification (etc. fluctuation and scare of rainwater), and population growth impose a huge pressure in their water supply (Bahadir,2016).

Water sustainability is the corner stone for this challenge

(Dhabi et al., 2020) إلى إمكانية المعالجة الحيوية للنفط الخام باستخدام سلالة Bacillus subtilis Al-Dhabi-130 المعزولة من الرواسب البحرية، حيث أظهرت النتائج أن B. subtilis تمكنت من إزالة 89% من النفط الخام.

14. دور بعض الدول العربية الأخرى في تكنولوجيا المعالجة البيولوجية

تعد دول الشرق الأوسط وشمال إفريقيا من أفقر الدول من حيث توفر المياه، إذ تتميز طبيعتها بالصحراء القاسية، ويعتبر المصدر الرئيسي لإمدادات المياه فيها هو المياه الجوفية. ومن أبرز التحديات في هذه المنطقة ندرة المياه، والتي تتجلى في زيادة الطلب على المياه للاستخدام المنزلي والزراعي. تستهلك الزراعة وحدها 70% من المياه المتاحة بسبب تقنيات الري غير الكافية. كما أن تغير المناخ المتمثل في ارتفاع درجات الحرارة وزيادة التصحر (مثل تذبذب وشح الأمطار) والنمو السكاني يفرض ضغوطاً كبيرة على إمدادات المياه (Bahadir,2016).

تعتبر استدامة المياه حجر الزاوية لمواجهة هذا التحدي، ويتجلى ذلك في أهداف التنمية

which is reflected by Sustainable Development Goals (SDG). For such, wastewater treatment programmes have been implemented by several countries in the Middle East and North Africa along with sea water desalination. Jordan is one of the poorest courtiers in MENA with regards to freshwater due to is limitation. The main source of water is groundwater, Dams. The level of rainwater in 2022 is 6192 MCM (MWI, 2022). 94% of rainwater is impact by evaporation, whereas only 4% is for groundwater recharge. The global benchmark for absolute water poverty is 500 m3 per capita. Jordan has reduced from 3,600 m3 per capita in 1940 to 61 m3 per capita in 2023 of fresh renewable water. This is considered as far less than the internationally recognised level.15% of Jordan water supply comes from 9 wastewater plant and this reducing the pressure on Jordan's freshwater resources.

GCC water share had reduced from 1200 MCM in 1950 to less than 200 MCM 2010. Wastewater is accounted for

المستدامة. لذلك، قامت عدة دول في الشرق الأوسط وشمال إفريقيا بتطبيق برامج لمعالجة مياه الصرف الصحي إلى جانب تحلية مياه البحر. وتعتبر الأردن من أفقر الدول في منطقة الشرق الأوسط وشمال إفريقيا فيما يتعلق بالمياه العذبة بسبب محدوديتها، حيث تعتمد على المياه الجوفية والسدود كمصادر رئيسية للمياه. بلغ مستوى مياه الأمطار في عام 2022 حوالي 6192 مليون متر مكعب (MWI، 2022)، حيث يتأثر 94% من مياه الأمطار بالتبخر، بينما 4% فقط تذهب لإعادة تغذية المياه الجوفية. ويعتبر المعيار العالمي للفقر المطلق في المياه 500 متر مكعب للفرد، وقد انخفض معدل الأردن من 3600 متر مكعب للفرد في عام 1940 إلى 61 متر مكعب للفرد في عام 2023 من المياه العذبة المتجددة، وهو أقل بكثير من المستوى المعترف به دولياً. يأتي 15% من إمدادات المياه في الأردن من 9 محطات لمعالجة مياه الصرف الصحي، مما يخفف الضغط على موارد المياه العذبة في الأردن.

انخفض نصيب دول مجلس التعاون الخليجي من المياه من 1200 مليون متر مكعب في عام 1950 إلى أقل من 200

only 3% whereas 78% and 19% comes from groundwater and desalinization, respectively. Agriculture is considered the main consumer accounting for 77%, 18% municipal, and 5% industrial. 4.0 billion m3 wastewater is annually collected and more than 300 treatment plant treat 73% of which (Qureshi, 2020).

With regards to North Africa, Egypt main source of fresh water is Nile river (55.5 BCM) which constitutes of 97% of its water supply. Its production of wastewater is 4.0 BCM which is discharged to main canal to be mixed with freshwater. The government's future planning to is to reuse 3.0 BCM (Belhassan,2022; Gado et al. 2020).

With regards to Libya, it relies heavily on ground water which constitutes 95% of its production. 2 % desalination and 2% surface water. Only 1% of wastewater is being treated from only 9 functional plants out 36 wastewater plants. Most of untreated wastewater is being disposed to Mediterranean Sea or into water canal systems (Alsadey,2020).

مليون متر مكعب في عام 2010. تمثل مياه الصرف الصحي 3% فقط، بينما 78% تأتي من المياه الجوفية و19% من التحلية. وتعد الزراعة المستهلك الرئيسي بنسبة 77%، تليها الاستخدام البلدي بنسبة 18%، ثم الصناعي بنسبة 5%. يتم جمع 4.0 مليار متر مكعب من مياه الصرف الصحي سنوياً ويتم معالجة أكثر من 73% منها عبر 300 محطة معالجة (Qureshi, 2020).

وفيما يتعلق بشمال إفريقيا، فإن المصدر الرئيسي للمياه العذبة في مصر هو نهر النيل (55.5 مليار متر مكعب)، والذي يمثل 97% من إمداداتها المائية. يبلغ إنتاج مصر من مياه الصرف الصحي 4.0 مليار متر مكعب يتم تصريفها في القنوات الرئيسية ليتم مزجها بالمياه العذبة. وتخطط الحكومة لإعادة استخدام 3.0 مليار متر مكعب مستقبلاً (Belhassan,2022; Gado et al. 2020).

أما بالنسبة لليبيا، فهي تعتمد بشكل كبير على المياه الجوفية التي تشكل 95% من إنتاجها، بالإضافة إلى 2% من التحلية و2% من المياه السطحية. يتم معالجة 1% فقط من مياه الصرف الصحي من خلال 9 محطات فقط من أصل 36 محطة لمعالجة

Tunisia has 86% of public sewage network connection which is considered to be high in comparison to its surrounding countries. 99% of which is being treated (Chaouali,2024).

15. The scientific and applied benefits

The environmental wastes in their liquid or solid forms are daily released into the environment in all developed and developing countries. Developing countries are looking at these wastes as a disaster. While the developed countries, on the contrary, are looking at these wastes as a probable source of wealth. Most developing countries are dealing with daily environmental wastes by the easy ways of incineration or landfilling. While the developed countries are depending on the strategy of collecting, sorting, and recycling of these wastes into beneficial products. No one can deny that some of the wastes are dangerous and can cause multiple risks to the humans and to the

مياه الصرف الصحي. ويتم التخلص من معظم مياه الصرف الصحي غير المعالجة في البحر الأبيض المتوسط أو في قنوات المياه (Alsadey,2020).
تمتلك تونس شبكة صرف صحي تغطي 86% من السكان، وتعد هذه النسبة مرتفعة مقارنة بالدول المجاورة، حيث يتم معالجة 99% من هذه المياه (Chaouali,2024).

15. الفوائد العلمية والتطبيقية

يتم إطلاق المخلفات البيئية بشكلها السائل أو الصلب يومياً إلى البيئة في جميع البلدان المتقدمة والنامية. وتنظر البلدان النامية إلى هذه المخلفات باعتبارها كارثة. بينما تنظر الدول المتقدمة، على العكس من ذلك، إلى هذه المخلفات كمصدر محتمل للثروة. تتعامل معظم البلدان النامية مع المخلفات البيئية اليومية بالطرق السهلة المتمثلة في الحرق أو دفن النفايات. بينما تعتمد الدول المتقدمة على استراتيجية جمع وفرز وإعادة تدوير هذه المخلفات إلى منتجات مفيدة. ولا يمكن لأحد أن ينكر أن بعض المخلفات خطيرة ويمكن أن تسبب مخاطر متعددة للإنسان والبيئة. ومع ذلك، فإن البعض الآخر يشكل مصادر للثراء، إذا تم استغلاله بشكل صحيح.

environment. However, some others are sources of rich, if properly exploited.

The scholars have globally paid a lot of attention for the environmental wastes in order to either mitigate their hazardous effects or to convert them into value-add products. Bioremediation using microorganisms has scientifically participated in both strategies. The microbes have been used to transform the toxic elements such as heavy metals into other less toxic forms through multiple mechanisms including bioleaching, biosorption, and bioaccumulation. They have been also participated in the biodegradation of toxic organic wastes such as hydrocarbons into less toxic forms in levels that should be safe for the surrounding environment. However, microorganisms have also been succeeded in the transform of some of these organic wastes into beneficial products such as enzymes, bioenergy, vitamins, organic acids, etc.

From a real point of view, a lot of the environmental wastes have been used as raw materials for the production of a lot of the aforementioned products. These organic environmental wastes are free, easy to collect, and daily available raw materials for a lot of industrial products. Most of the food and kitchen wastes are daily thrown into the environment with huge amounts. Such wastes are rich with organic molecules such as carbohydrates, proteins, and lipids that can be used as raw materials for the microbial production of enzymes including amylases, proteases, and lipases with the minimum cost. However, the carbohydrate or cellulosic parts of food and kitchen wastes as well as the agricultural wastes can be used for the production of fuels such as bioethanol or biobutanol. In recent studies, the food and kitchen wastes have been used for the production of biocellulose which is gaining a lot of interest and applications in the biomedical, cosmetic, agricultural, and environmental fields.

On the other hand, the contaminated domestic, agricultural, and industrial wastewater are another

الكثير من المنتجات المذكورة أعلاه. وهذه المخلفات البيئية العضوية مجانية، وسهلة التجميع، ومواد أولية متوافرة يومياً للكثير من المنتجات الصناعية. يتم إلقاء معظم مخلفات الطعام والمطبخ يوميًا في البيئة بكميات هائلة. هذه المخلفات غنية بالجزيئات العضوية مثل الكربوهيدرات والبروتينات والدهون التي يمكن استخدامها كمواد خام لإنتاج الإنزيمات الميكروبية بما في ذلك الأميليز والبروتياز والليباز بأقل تكلفة. ومع ذلك، يمكن استخدام الأجزاء الكربوهيدراتية أو السليولوزية من نفايات الطعام والمطبخ وكذلك النفايات الزراعية لإنتاج الوقود مثل الإيثانول الحيوي أو البيوتانول الحيوي. في الدراسات الحديثة، تم استخدام مخلفات الطعام والمطبخ لإنتاج السليلوز الحيوي الذي يحظى بالكثير من الاهتمام والتطبيقات في المجالات الطبية الحيوية والتجميلية والزراعية والبيئية.

من ناحية أخرى، تعتبر مياه الصرف الصحي المنزلية والزراعية والصناعية الملوثة مصدرا غنيا آخر، خاصة مع أزمة شح المياه المتوقعة. ويمكن معالجة مصادر المياه هذه ميكروبياً من أجل معالجة الكثير

wealthy source, especially with the predicted water scarcity crisis. These water sources can be microbially treated in order to remediate a lot of the water inhabiting contaminants to be reused for the irrigation purposes, or incorporated in industrial steps during the industrial cycles of production, according to the quality of treated water.

16. Conclusion

The bioremediation of environmental wastes using microbes has been succeeded in remediating both of organic and inorganic pollutants. The microbes were able to use different strategies such as bioleaching, biosorption, bioaccumulation, enzymatic oxidation or reduction, in addition to other techniques to get rid of these pollutants in a safe way. They succeeded to convert the surrounding pollutants into non or less harmful ones, and in most cases converting them into value-added products such as organic acids, bioenergy, and other biomolecules that have wide applications in medical, industrial, and environmental fields.

من الملوثات الموجودة في المياه لإعادة استخدامها لأغراض الري، أو دمجها في الخطوات الصناعية خلال دورات الإنتاج الصناعية، وفقاً لنوعية المياه المعالجة.

16. الخاتمة

لقد نجحت المعالجة الحيوية للنفايات البيئية باستخدام الميكروبات في معالجة كل من الملوثات العضوية وغير العضوية. وتمكنت الميكروبات من استخدام استراتيجيات مختلفة مثل الترشيح الحيوي، والامتصاص الحيوي، والتراكم الحيوي، والأكسدة الأنزيمية أو الاختزال، بالإضافة إلى تقنيات أخرى للتخلص من هذه الملوثات بطريقة آمنة. لقد نجحوا في تحويل الملوثات المحيطة إلى ملوثات غير ضارة أو أقل ضرراً، وفي معظم الحالات تحويلها إلى منتجات ذات قيمة عالية مثل الأحماض العضوية والطاقة الحيوية وغيرها من الجزيئات الحيوية التي لها تطبيقات واسعة في المجالات الطبية والصناعية والبيئية.

17. References   17. المراجع

Abdel-Azeem AM, Hasan GA, Mohesien MT (2020) Biodegradation of Agricultural Wastes by Chaetomium Species. Recent Developments on Genus Chaetomium:301-341

Abdel-Razek A, El-Sheikh H, Suleiman W, Taha TH, Mohamed M (2020) Bioelimination of phenanthrene using degrading bacteria isolated from petroleum soil: safe approach. Desalination and water treatment 181:131-140

Abdel-Razek MA, Abozeid AM, Eltholth MM, Abouelenien FA, El-Midany SA, Moustafa NY, Mohamed RA (2019) Bioremediation of a pesticide and selected heavy metals in wastewater from various sources using a consortium of microalgae and cyanobacteria. Slov Vet 56 (Suppl 22):61-73

Abdel-Razik MA, Azmy AF, Khairalla AS, AbdelGhani S (2020) Metal bioremediation potential of the halophilic bacterium, Halomonas sp. strain WQL9 isolated from Lake Qarun, Egypt. The Egyptian Journal of Aquatic Research 46 (1):19-25

Abdelaal M, Mashaly IA, Srour DS, Dakhil MA, El-Liethy MA, El-Keblawy A, El-Barougy RF, Halmy MWA, El-Sherbeny GA (2021) Phytoremediation perspectives of seven aquatic macrophytes for removal of heavy metals from polluted drains in the Nile Delta of Egypt. Biology 10 (6):560

ABDULMOHSEN, K. A., ALIMI, F. R., MECHI, L., KAA, A. A., MOHAMMAD, A. E. & KHAN, M. W. A. 2023. Optimization of Pseudomonas aeruginosa isolated for bioremediation from Ha'il region of Saudi Arabia. *Bioinformation,* 19**,** 893.

Abdulyekeen KA, Giwa SO, Ibrahim AA (2022) Bioremediation of used motor oil-contaminated soil using animal dung as stimulants. In: Animal Manure: Agricultural and Biotechnological Applications. Springer, pp 185-215

Abu-Saied M, Elnouby M, Taha T, El-Shafeey M, G. Alshehri A, Alamri S, Alghamdi H, Shati A, Alrumman S, Al-Kahtani M (2020) Potential decontamination of drinking water pathogens through k-carrageenan integrated green bottle fly bio-synthesized silver nanoparticles. Molecules 25 (8):1936

Adegboye MF, Ojuederie OB, Talia PM, Babalola OO (2021) Bioprospecting of microbial strains for biofuel production: metabolic engineering, applications, and challenges. Biotechnology for biofuels 14 (1):1-21

AH Ibrahim H, M Bassiouny Beliah M, M Farag A, MD ElAhwany A, A Sabry S (2020) Properties and aerogel applications of a marine algal origin biocellulose produced by the immobilized Gluconacetobacter xylinus ATCC 10245. Egyptian Journal of Aquatic Biology and Fisheries 24 (7-Special issue):51-71

Ahanchi M, Jafary T, Yeneneh AM, Rupani PF, Shafizadeh A, Shahbeik H, Pan J, Tabatabaei M, Aghbashlo M (2022) Review on waste biomass valorization and power management systems for microbial fuel cell application. Journal of Cleaner Production:134994

Ahirwar R, Tripathi AK (2021) E-waste management: A review of recycling process, environmental and occupational health hazards, and potential solutions. Environmental Nanotechnology, Monitoring & Management 15:100409

Ahmed M, Mavukkandy MO, Giwa A, Elektorowicz M, Katsou E, Khelifi O, Naddeo V, Hasan SW (2022) Recent developments in hazardous pollutants removal from wastewater and water reuse within a circular economy. NPJ Clean Water 5 (1):12

AL-DHABI, N. A., ESMAIL, G. A. & VALAN ARASU, M. 2020. Enhanced production of biosurfactant from Bacillus subtilis strain Al-Dhabi-130 under solid-state fermentation using date molasses from Saudi Arabia for bioremediation of crude-oil-contaminated soils. *International Journal of Environmental Research and Public Health,* 17**,** 8446.

Al-Tohamy R, Kenawy E-R, Sun J, Ali SS (2020) Performance of a newly isolated salt-tolerant yeast strain Sterigmatomyces halophilus SSA-1575 for azo dye decolorization and detoxification. Frontiers in Microbiology 11:1163

Al-Tohamy R, Sun J, Khalil MA, Kornaros M, Ali SS (2021) Wood-feeding termite gut symbionts as an obscure yet promising source of novel manganese peroxidase-producing oleaginous yeasts intended for azo dye decolorization and biodiesel production. Biotechnology for biofuels 14 (1):1-27

Alao MB, Adebayo EA (2022) Fungi as veritable tool in bioremediation of polycyclic aromatic hydrocarbons-polluted wastewater. Journal of Basic Microbiology 62 (3-4):223-244

Albelda-Berenguer M, Monachon M, Joseph E (2019) Siderophores: From natural roles to potential applications. Advances in applied microbiology 106:193-225

Ali SS, Al-Tohamy R, Khalil MA, Ho S-H, Fu Y, Sun J (2022a) Exploring the potential of a newly constructed manganese peroxidase-producing yeast consortium for tolerating lignin degradation inhibitors while simultaneously decolorizing and detoxifying textile azo dye wastewater. Bioresource Technology 351:126861

Ali SS, Al-Tohamy R, Mohamed TM, Mahmoud YA-G, Ruiz HA, Sun L, Sun J (2022b) Could termites be hiding a goldmine of obscure yet promising yeasts for energy crisis solutions based on aromatic wastes? A critical state-of-the-art review. Biotechnology for Biofuels and Bioproducts 15 (1):1-40

Ali SS, Elsamahy T, Al-Tohamy R, Zhu D, Mahmoud YA-G, Koutra E, Metwally MA, Kornaros M, Sun J (2021a) Plastic wastes biodegradation: Mechanisms, challenges and future prospects. Science of the Total Environment 780:146590

Ali SS, Sun J, Koutra E, El-Zawawy N, Elsamahy T, El-Shetehy M (2021b) Construction of a novel cold-adapted oleaginous yeast consortium valued for textile azo dye wastewater processing and biorefinery. Fuel 285:119050

Alsadey, S. and Mansour, O., 2020. Wastewater treatment plants in Libya: Challenges and future prospects. International Journal of Environmental Planning and Management, 6(3), pp.76-80.

Ambaye TG, Vaccari M, Franzetti A, Prasad S, Formicola F, Rosatelli A, Hassani A, Aminabhavi TM, Rtimi S (2023) Microbial electrochemical bioremediation of petroleum hydrocarbons (PHCs) pollution: Recent advances and outlook. Chemical Engineering Journal 452:139372

Amin SK, Roushdy MH, Abdallah HA, Moustafa AF, Abadir MF (2020) Preparation and characterization of ceramic nanofiltration membrane prepared from hazardous industrial waste. International Journal of Applied Ceramic Technology 17 (1):162-174

Amobonye A, Bhagwat P, Singh S, Pillai S (2021) Plastic biodegradation: Frontline microbes and their enzymes. Science of the Total Environment 759:143536

Angajala G, Aruna V, Pavan P, Reddy PG (2022) Biocatalytic one pot three component approach: Facile synthesis, characterization, molecular modelling and hypoglycemic studies of new thiazolidinedione festooned quinoline analogues catalyzed by alkaline protease from Aspergillus niger. Bioorganic Chemistry 119:105533

Anichebe CO, Okoye EL, Onochie CC (2019) Comparative study on single cell protein (SCP) production by Trichoderma viride from pineapple wastes and banana peels. International Journal of Research Publication, Forthcoming

Ansari AA, Naeem M, Gill SS, AlZuaibr FM (2020) Phytoremediation of contaminated waters: An eco-friendly technology based on aquatic macrophytes application. The Egyptian Journal of Aquatic Research 46 (4):371-376

Attrah M, Elmanadely A, Akter D, Rene ER (2022) A Review on Medical Waste Management: Treatment, Recycling, and Disposal Options. Environments 9 (11):146

Ayangbenro AS, Babalola OO, Aremu OS (2019) Bioflocculant production and heavy metal sorption by metal resistant bacterial isolates from gold mining soil. Chemosphere 231:113-120

Ayilara MS, Babalola OO (2023) Bioremediation of environmental wastes: the role of microorganisms. Frontiers in Agronomy 5:1183691

Babalola OO, Aremu BR, Ayangbenro AS (2019) Draft genome sequence of heavy metal-resistant Bacillus cereus NWUAB01. Microbiology Resource Announcements 8 (7):10.1128/mra. 01706-01718

Bahadir, M., Aydin, M.E., Aydin, S., Beduk, F. and Batarseh, M., 2016. Wastewater reuse in Middle East Countries—A review of prospects and challenges. Fresenius Environmental Bulletin, 25(5), pp.1284-1304.

Bai L, Zhang Q, Wang C, Yao X, Zhang H, Jiang H (2019) Effects of natural dissolved organic matter on the complexation and biodegradation of 17α-ethinylestradiol in freshwater lakes. Environmental Pollution 246:782-789

Baltazar MdPG, Gracioso LH, Avanzi IR, Karolski B, Tenório JAS, do Nascimento CAO, Perpetuo EA (2019) Copper biosorption by Rhodococcus erythropolis isolated from the Sossego Mine–PA–Brazil. Journal of Materials Research and Technology 8 (1):475-483

Bardhan P, Gupta K, Mandal M (2019) Microbes as bio-resource for sustainable production of biofuels and other bioenergy products. In: New and future developments in microbial biotechnology and bioengineering. Elsevier, pp 205-222

Barrera EL, Hertel T (2021) Global food waste across the income spectrum: Implications for food prices, production and resource use. Food Policy 98:101874

Behr M, Guerriero G, Grima-Pettenati J, Baucher M (2019) A molecular blueprint of lignin repression. Trends in plant science 24 (11):1052-1064

Belhassan, K., 2022. Managing drought and water stress in Northern Africa. In Arid Environment-Perspectives, Challenges and Management. IntechOpen.

Bhandari G, Gupta S, Chaudhary P, Chaudhary S, Gangola S (2023) Bioleaching: A Sustainable Resource Recovery Strategy for Urban Mining of E-waste. In: Microbial Technology for Sustainable E-waste Management. Springer, pp 157-175

Bhandari S, Poudel DK, Marahatha R, Dawadi S, Khadayat K, Phuyal S, Shrestha S, Gaire S, Basnet K, Khadka U (2021) Microbial enzymes used in bioremediation. Journal of Chemistry 2021:1-17

Bhatt P, Bhandari G, Bhatt K, Maithani D, Mishra S, Gangola S, Bhatt R, Huang Y, Chen S (2021a) Plasmid-mediated catabolism for the removal of xenobiotics from the environment. Journal of Hazardous Materials 420:126618

Bhatt P, Bhatt K, Chen W-J, Huang Y, Xiao Y, Wu S, Lei Q, Zhong J, Zhu X, Chen S (2023) Bioremediation potential of laccase for catalysis of glyphosate, isoproturon, lignin, and parathion: Molecular docking, dynamics, and simulation. Journal of Hazardous Materials 443:130319

Bhatt P, Tiwari M, Parmarick P, Bhatt K, Gangola S, Adnan M, Singh Y, Bilal M, Ahmed S, Chen S (2022) Insights into the catalytic mechanism of ligninolytic peroxidase and laccase in lignin degradation. Bioremediation Journal 26 (4):281-291

Bhatt P, Verma A, Gangola S, Bhandari G, Chen S (2021b) Microbial glycoconjugates in organic pollutant bioremediation: recent advances and applications. Microbial Cell Factories 20 (1):1-18

Bolade OP, Williams AB, Benson NU (2020) Green synthesis of iron-based nanomaterials for environmental remediation: A review. Environmental Nanotechnology, Monitoring & Management 13:100279

Brusseau M (2019) Soil and groundwater remediation. In: Environmental and pollution science. Elsevier, pp 329-354

Bui NK, Nguyen TL, Phan KD, Nguyen AT Legal framework for recycling domestic solid waste in Vietnam: situation and recommendation. In: E3S Web of Conferences, 2020. EDP Sciences, p 11009

Cesaro A, Belgiorno V, Gorrasi G, Viscusi G, Vaccari M, Vinti G, Jandric A, Dias MI, Hursthouse A, Salhofer S (2019) A relative risk assessment of the open burning of WEEE. Environmental Science and Pollution Research 26:11042-11052

Chandra P, Enespa, Singh R, Arora PK (2020) Microbial lipases and their industrial applications: a comprehensive review. Microbial cell factories 19:1-42

Chandra V, Arpita K, Yadav P, Raghuvanshi V, Yadav A, Prajapati S (2023) Environmental Biotechnology for

Medical Waste Management: A Review of Current Practices and Future Directions.

Chaouali, S., Dos Muchangos, L.S., Ito, L. and Tokai, A., 2024. Assessment of the Environmental Impacts of Wastewater Treatment in Tunisia. Journal of Water and Environment Technology, 22(2), pp.61-74.

Chen F, Zeng S, Luo Z, Ma J, Zhu Q, Zhang S (2020) A novel MBBR–MFC integrated system for high-strength pulp/paper wastewater treatment and bioelectricity generation. Separation Science and Technology 55 (14):2490-2499

Cheng S, Li N, Jiang L, Li Y, Xu B, Zhou W (2019) Biodegradation of metal complex Naphthol Green B and formation of iron–sulfur nanoparticles by marine bacterium Pseudoalteromonas sp CF10-13. Bioresource technology 273:49-55

Chhaya, Raychoudhury T, Prajapati SK (2020) Bioremediation of pharmaceuticals in water and wastewater. Microbial bioremediation & biodegradation:425-446

Chilakamarry CR, Sakinah AM, Zularisam A, Sirohi R, Khilji IA, Ahmad N, Pandey A (2022) Advances in solid-state fermentation for bioconversion of agricultural wastes to value-added products: Opportunities and challenges. Bioresource technology 343:126065

Dabbagh F, Moradpour Z, Ghasemian A (2019) Microbial products and biotechnological applications thereof: Proteins, enzymes, secondary metabolites, and valuable chemicals. Microbial Interventions in Agriculture and Environment: Volume 3: Soil and Crop Health Management:385-432

Dabe SJ, Prasad PJ, Vaidya A, Purohit H (2019) Technological pathways for bioenergy generation from municipal solid waste: Renewable energy option. Environmental Progress & Sustainable Energy 38 (2):654-671

Dar MA, Syed R, Pawar KD, Dhole NP, Xie R, Pandit RS, Sun J (2022) Evaluation and characterization of the cellulolytic bacterium, Bacillus pumilus SL8 isolated from the gut of oriental leafworm Spodoptera litura: An assessment of its potential value for lignocellulose bioconversion. Environmental Technology & Innovation 27:102459

Darwesh O, Shalapy M, Abo-Zeid A, Mahmoud Y (2021) Nano-bioremediation of municipal wastewater using myco-synthesized iron nanoparticles. Egyptian Journal of Chemistry 64 (5):2499-2507

Darwesh OM, Matter IA, Eida MF (2019) Development of peroxidase enzyme immobilized magnetic nanoparticles for bioremediation of textile wastewater dye. Journal of Environmental Chemical Engineering 7 (1):102805

Das AK, Islam MN, Billah MM, Sarker A (2021) COVID-19 pandemic and healthcare solid waste management strategy–A mini-review. Science of the Total Environment 778:146220

Das S, Lee S-H, Kumar P, Kim K-H, Lee SS, Bhattacharya SS (2019) Solid waste management: Scope and the challenge of sustainability. Journal of cleaner production 228:658-678

Dave S, Das J (2021) Role of microbial enzymes for biodegradation and bioremediation of environmental pollutants: challenges and future prospects. Bioremediation for Environmental Sustainability:325-346

Debnath B, Majumdar M, Bhowmik M, Bhowmik KL, Debnath A, Roy DN (2020) The effective adsorption of tetracycline onto zirconia nanoparticles synthesized by novel microbial green technology. Journal of environmental management 261:110235

Deshpande B, Agrawal P, Yenkie M, Dhoble S (2020) Prospective of nanotechnology in degradation of waste water: A new challenges. Nano-Structures & Nano-Objects 22:100442

Dong X, Zhang YQ (2021) An insight on egg white: From most common functional food to biomaterial application. Journal of Biomedical Materials Research Part B: Applied Biomaterials 109 (7):1045-1058

Duan M, Zhang Y, Zhou B, Qin Z, Wu J, Wang Q, Yin Y (2020) Effects of Bacillus subtilis on carbon components and microbial functional metabolism during cow manure–straw composting. Bioresource Technology 303:122868

Ebah E, Yange I, Ohie I, Inya O (2022) Application of Genetically Modified Organisms in Waste Management–A Review. Stamford Journal of Microbiology 12 (1):15-20

El-Badry M, Elbarbary T, Abdel-Fatah Y, Abdel-Halim S, Sharada H, Ibrahim IA (2021) Role of Actinomycete sp. in Bio-extraction of Copper from Electronic Waste.

El Hammoudani Y, Dimane F, El Ouarghi H (2021) Removal efficiency of heavy metals by a biological wastewater treatment plant and their potential risks to human health. Environmental Engineering and Management Journal 20 (6):995-1002

Elshamy MM, Heikal YM, Bonanomi G (2019) Phytoremediation efficiency of Portulaca oleracea L. naturally growing in some industrial sites, Dakahlia District, Egypt. Chemosphere 225:678-687

Elzakey EM, El-Sabbagh SM, Eldeen EE-SN, Adss IA-A, Nassar AMK (2023) Bioremediation of chlorpyrifos residues using some indigenous species of bacteria and fungi in wastewater. Environmental Monitoring and Assessment 195 (6):779

Espinosa-Ortiz EJ, Rene ER, Gerlach R (2022) Potential use of fungal-bacterial co-cultures for the removal of organic pollutants. Critical reviews in biotechnology 42 (3):361-383

Fadzli FS, Bhawani SA, Adam Mohammad RE (2021) Microbial fuel cell: recent developments in organic substrate use and bacterial electrode interaction. Journal of Chemistry 2021:1-16

Fashola MO, Ngole-Jeme VM, Babalola OO (2020) Heavy metal immobilization potential of indigenous bacteria isolated from gold mine tailings. International Journal of Environmental Research 14:71-86

Fierascu RC, Sieniawska E, Ortan A, Fierascu I, Xiao J (2020) Fruits by-products–A source of valuable active principles. A short review. Frontiers in bioengineering and biotechnology 8:319

Fierer N, Wood SA, de Mesquita CPB (2021) How microbes can, and cannot, be used to assess soil health. Soil Biology and Biochemistry 153:108111

Forti V, Balde CP, Kuehr R, Bel G (2020) The Global E-waste Monitor 2020: Quantities, flows and the circular economy potential.

Gaber SE, Hussain MT, Jahin HS (2020) Bioremediation of diazinon pesticide from aqueous solution by fungal-strains isolated from wastewater. World J Chem 15 (1):15-23

Gado, T.A. and El-Agha, D.E., 2020. Feasibility of rainwater harvesting for sustainable water management in urban areas of Egypt. Environmental Science and Pollution Research, 27(26), pp.32304-32317.

Gangola S, Joshi S, Kumar S, Pandey S (2019) Comparative analysis of fungal and bacterial enzymes in biodegradation of xenobiotic compounds. Smart Bioremediation Technologies. Elsevier,

Gao L-L, Lu Y-C, Zhang J-L, Li J, Zhang J-D (2019) Biotreatment of restaurant wastewater with an oily high concentration by newly isolated bacteria from oily sludge. World Journal of Microbiology and Biotechnology 35:1-11

Ghorai P, Ghosh D (2022a) Ameliorating the performance of NPK biofertilizers to attain sustainable agriculture with special emphasis on bioengineering. Bioresource Technology Reports 19:101117

Ghorai P, Ghosh D (2022b) Sustainable approach for insoluble phosphate recycling from wastewater effluents. In: Environmental Degradation: Monitoring, Assessment and Treatment Technologies. Springer, pp 77-86

Ghosh D, Ghorai P, Sarkar S, Maiti KS, Hansda SR, Das P (2023) Microbial assemblage for solid waste bioremediation and valorization with an essence of bioengineering. Environmental Science and Pollution Research 30 (7):16797-16816

Goda DA, Diab MA, El-Gendi H, Kamoun EA, Soliman NA, Saleh AK (2022) Fabrication of biodegradable chicken feathers into ecofriendly-functionalized biomaterials: characterization and bio-assessment study. Scientific Reports 12 (1):18340

Golbaz S, Zamanzadeh MZ, Pasalari H, Farzadkia M (2021) Assessment of co-composting of sewage sludge, woodchips, and sawdust: feedstock quality and design and compilation of computational model. Environmental Science and Pollution Research 28:12414-12427

Govarthanan M, Jeon C-H, Jeon Y-H, Kwon J-H, Bae H, Kim W (2020) Non-toxic nano approach for wastewater treatment using Chlorella vulgaris exopolysaccharides immobilized in iron-magnetic nanoparticles. International Journal of Biological Macromolecules 162:1241-1249

Gricajeva A, Nadda AK, Gudiukaite R (2022) Insights into polyester plastic biodegradation by carboxyl ester hydrolases. Journal of Chemical Technology & Biotechnology 97 (2):359-380

Guo J, Liu X, Zhang X, Wu J, Chai C, Ma D, Chen Q, Xiang D, Ge W (2019) Immobilized lignin peroxidase on Fe_3O_4@ SiO_2@ polydopamine nanoparticles for degradation of organic pollutants. International journal of biological macromolecules 138:433-440

Gupta GK, Shukla P (2020) Insights into the resources generation from pulp and paper industry wastes: challenges, perspectives and innovations. Bioresource technology 297:122496

Gupta S, Dangi L, Patra JK, Rani R (2021) Application of Enzymes in Bioremediation of Contaminated Hydrosphere and Soil Environment. Bioprospecting of Enzymes in Industry, Healthcare and Sustainable Environment:1-28

Gustafsson U, O'Connell R, Draper A, Tonner A (2019) What is Food?: Researching a Topic with Many Meanings. Routledge,

Habib A, Bhatti HN, Iqbal M (2020) Metallurgical processing strategies for metals recovery from industrial slags. Zeitschrift für Physikalische Chemie 234 (2):201-231

Han P, Teo WZ, Yew WS (2022) Biologically engineered microbes for bioremediation of electronic waste: Wayposts, challenges and future directions. Engineering Biology 6 (1):23-34

HASHEM, A. 2007. Bioremediation of petroleum contaminated soils in the Arabian Gulf region: a review. *Science,* 19.

Hofrichter M, Kellner H, Herzog R, Karich A, Liers C, Scheibner K, Kimani VW, Ullrich R (2020) Fungal peroxygenases: A phylogenetically old superfamily of heme enzymes with promiscuity for oxygen transfer reactions. Grand challenges in fungal biotechnology:369-403

Humer D, Ebner J, Spadiut O (2020) Scalable high-performance production of recombinant horseradish peroxidase from E. coli inclusion bodies. International Journal of Molecular Sciences 21 (13):4625

Hung Y-H, Liu T-Y, Chen H-Y (2019) Renewable coffee waste-derived porous carbons as anode materials for high-performance sustainable microbial fuel cells. ACS Sustainable Chemistry & Engineering 7 (20):16991-16999

Hussain A, Rehman F, Rafeeq H, Waqas M, Asghar A, Afsheen N, Rahdar A, Bilal M, Iqbal HM (2022) In-situ, Ex-situ, and nano-remediation strategies to treat polluted soil, water, and air–A review. Chemosphere 289:133252

Hussein SH, Qurbani K, Ahmed SK, Tawfeeq W, Hassan M (2023) Bioremediation of heavy metals in contaminated environments using Comamonas species: A narrative review. Bioresource Technology Reports:101711

Huy ND, Le NTM, Chew KW, Park S-M, Show PL (2021) Characterization of a recombinant laccase from Fusarium oxysporum HUIB02 for biochemical application on dyes removal. Biochemical Engineering Journal 168:107958

Ieropoulos I, Greenman J (2023) The future role of MFCs in biomass energy. Frontiers in Energy Research 11:1108389

Jaiswal S, Sharma B, Shukla P (2020) Integrated approaches in microbial degradation of plastics. Environmental Technology & Innovation 17:100567

Jamal MT, Pugazhendi A, Jeyakumar RB (2020) Application of halophiles in air cathode MFC for seafood industrial wastewater treatment and energy production under high saline condition. Environmental Technology & Innovation 20:101119

JeyaSundar PGSA, Ali A, Zhang Z (2020) Waste treatment approaches for environmental sustainability. In: Microorganisms for sustainable environment and health. Elsevier, pp 119-135

Jha H (2019) Fungal diversity and enzymes involved in lignin degradation. Mycodegradation of Lignocelluloses:35-49

Jie H, Khan I, Alharthi M, Zafar MW, Saeed A (2023) Sustainable energy policy, socio-economic development, and ecological footprint: The economic significance of natural resources, population growth, and industrial development. Utilities Policy 81:101490

Jin X, Song J, Liu G-Q (2020) Bioethanol production from rice straw through an enzymatic route mediated by enzymes developed in-house from Aspergillus fumigatus. Energy 190:116395

JONES, D. A., HAYES, M., KRUPP, F., SABATINI, G., WATT, I. & WEISHAR, L. 2008. The impact of the Gulf War (1990–91) oil release upon the intertidal Gulf coast line of Saudi Arabia and subsequent recovery. *Protecting the Gulf's marine ecosystems from pollution.* Springer.

Kalayu G (2019) Phosphate solubilizing microorganisms: promising approach as biofertilizers. International Journal of Agronomy 2019:1-7

Kapahi M, Sachdeva S (2019) Bioremediation options for heavy metal pollution. Journal of health and pollution 9 (24):191203

Kaur M, Sodhi H (2022) Genetically Engineered Microorganisms for Bioremediation Processes. In: Microbial Bioremediation: Sustainable Management of Environmental Contamination. Springer, pp 91-107

Khaliq N (2023) Microbial enzymes as a robust process to mitigate pollutants of environmental concern. In: Microbial Biomolecules. Elsevier, pp 241-267

khamis Soliman N, Moustafa AF, Aboud AA, Halim KSA (2019) Effective utilization of Moringa seeds waste as a new green environmental adsorbent for removal of industrial toxic dyes. Journal of Materials Research and Technology 8 (2):1798-1808

Kim T, Yeo J, Yang Y, Kang S, Paek Y, Kwon JK, Jang JK (2019) Boosting voltage without electrochemical degradation using energy-harvesting circuits and power management system-coupled multiple microbial fuel cells. Journal of Power Sources 410:171-178

KOÇAK E, İKİZOĞLU B (2020) Types of waste in the context of waste management and general overview of waste disposal in Turkey. International Journal of Agriculture Environment and Food Sciences 4 (4):520-527

Koutra E, Mastropetros SG, Ali SS, Tsigkou K, Kornaros M (2021) Assessing the potential of Chlorella vulgaris for valorization of liquid digestates from agro-industrial and municipal organic wastes in a biorefinery approach. Journal of Cleaner Production 280:124352

Kumar H, Bhardwaj K, Sharma R, Nepovimova E, Kuča K, Dhanjal DS, Verma R, Bhardwaj P, Sharma S, Kumar D (2020a) Fruit and vegetable peels: Utilization of high value horticultural waste in novel industrial applications. Molecules 25 (12):2812

Kumar M, Borah P, Devi P (2020b) Priority and emerging pollutants in water. In: Inorganic Pollutants in Water. Elsevier, pp 33-49

Kumar V, Agrawal S, Bhat SA, Américo-Pinheiro JHP, Shahi SK, Kumar S (2022) Environmental impact, health hazards, and plant-microbes synergism in remediation of emerging contaminants. Cleaner chemical engineering 2:100030

Landberg T, Greger M (2022) Phytoremediation Using Willow in Industrial Contaminated Soil. Sustainability 14 (14):8449

Lee SM, Lee D (2022) Effective Medical waste management for sustainable green healthcare. International Journal of Environmental Research and Public Health 19 (22):14820

Lenzen M, Malik A, Li M, Fry J, Weisz H, Pichler P-P, Chaves LSM, Capon A, Pencheon D (2020) The environmental footprint of health care: a global assessment. The Lancet Planetary Health 4 (7):e271-e279

Li H, He Y, Yan Z, Yang Z, Tian F, Liu X, Wu Z (2023) Insight into the microbial mechanisms for the improvement of spent mushroom substrate composting efficiency driven by phosphate-solubilizing Bacillus subtilis. Journal of Environmental Management 336:117561

Li Z, Chen Z, Zhu Q, Song J, Li S, Liu X (2020) Improved performance of immobilized laccase on Fe3O4@ C-Cu2+ nanoparticles and its application for biodegradation of dyes. Journal of Hazardous Materials 399:123088

Liu S, Xu X, Kang Y, Xiao Y, Liu H (2020) Degradation and detoxification of azo dyes with recombinant ligninolytic enzymes from Aspergillus sp. with secretory overexpression in Pichia pastoris. Royal Society Open Science 7 (9):200688

Liu Z, Tran K-Q (2021) A review on disposal and utilization of phytoremediation plants containing heavy metals. Ecotoxicology and Environmental Safety 226:112821

Long C, Jiang Z, Shangguan J, Qing T, Zhang P, Feng B (2021) Applications of carbon dots in environmental pollution control: A review. Chemical Engineering Journal 406:126848

López-Fernández J, Benaiges MD, Valero F (2021) Second-and third-generation biodiesel production with immobilised recombinant Rhizopus oryzae lipase: Influence of the support, substrate acidity and bioprocess scale-up. Bioresource Technology 334:125233

Mahanty S, Chatterjee S, Ghosh S, Tudu P, Gaine T, Bakshi M, Das S, Das P, Bhattacharyya S, Bandyopadhyay S (2020) Synergistic approach towards the sustainable management of heavy metals in wastewater using mycosynthesized iron oxide nanoparticles: Biofabrication, adsorptive dynamics and chemometric modeling study. Journal of Water Process Engineering 37:101426

Mahapatra S, Yadav R, Ramakrishna W (2022) Bacillus subtilis impact on plant growth, soil health and environment: Dr. Jekyll and Mr. Hyde. Journal of Applied Microbiology 132 (5):3543-3562

Mahmoud GA-E (2021) Microbial scavenging of heavy metals using bioremediation strategies. Rhizobiont in bioremediation of hazardous waste:265-289

Maity JP, Chen G-S, Huang Y-H, Sun A-C, Chen C-Y (2019) Ecofriendly heavy metal stabilization: microbial induced mineral precipitation (MIMP) and biomineralization for heavy metals within the contaminated soil by indigenous bacteria. Geomicrobiology Journal 36 (7):612-623

Malakar N, Mitra S, Toppo P, Mathur P (2020) Understanding the functional attributes of different microbial enzymes in bioremediation.

Mandeep, Shukla P (2020) Microbial nanotechnology for bioremediation of industrial wastewater. Frontiers in Microbiology 11:590631

Mangla D, Abbasi A, Aggarwal S, Manzoor K, Ahmad S, Ikram S (2019) Effective removal of "non-biodegradable" pollutants from contaminated water. Metal oxide-based photocatalyst for the degradation of organic pollutants in water:159

Manisalidis I, Stavropoulou E, Stavropoulos A, Bezirtzoglou E (2020) Environmental and health impacts of air pollution: a review. Frontiers in public health 8:14

Mansy AE, El Desouky EA, Taha TH, Abu-Saied M, El-Gendi H, Amer RA, Tian Z-Y (2024) Sustainable production of bioethanol from office paper waste and its purification via blended polymeric membrane. Energy Conversion and Management 299:117855

Manzoor J, Sharma M (2019) Impact of biomedical waste on environment and human health. Environmental Claims Journal 31 (4):311-334

McCarty NS, Ledesma-Amaro R (2019) Synthetic biology tools to engineer microbial communities for biotechnology. Trends in biotechnology 37 (2):181-197

Medfu Tarekegn M, Zewdu Salilih F, Ishetu AI (2020) Microbes used as a tool for bioremediation of heavy metal from the environment. Cogent Food & Agriculture 6 (1):1783174

Melanouri E-M, Dedousi M, Diamantopoulou P (2022) Cultivating Pleurotus ostreatus and Pleurotus eryngii mushroom strains on agro-industrial residues in solid-state fermentation. Part I: Screening for growth, endoglucanase, laccase and biomass production in the colonization phase. Carbon Resources Conversion 5 (1):61-70

Melnichuk N, Braia MJ, Anselmi PA, Meini M-R, Romanini D (2020) Valorization of two agroindustrial wastes to produce alpha-amylase enzyme from Aspergillus oryzae by solid-state fermentation. Waste Management 106:155-161

Mohanan N, Montazer Z, Sharma PK, Levin DB (2020) Microbial and enzymatic degradation of synthetic plastics. Frontiers in Microbiology 11:580709

Mohapatra B, Dhamale T, Saha BK, Phale PS (2022) Microbial degradation of aromatic pollutants: metabolic routes, pathway diversity, and strategies for bioremediation. In: Microbial biodegradation and bioremediation. Elsevier, pp 365-394

Mostafa YS, Alamri SA, Alrumman SA, Taha TH, Hashem M, Moustafa M, Fahmy LI (2021) Biosynthesis of raw starch degrading β-cyclodextrin glycosyltransferase by immobilized cells of Bacillus licheniformis using potato wastewater. Biocell 45 (6):1661

MOUSTAFA, A. 2016. Bioremediation of oil spill in Kingdom of Saudi Arabia by using fungi isolated from polluted soils. *International Journal of Current Microbiology and Applied Sciences,* 5, 680-91.

Moustafa M, Abu-Saied M, Taha T, Elnouby M, El-Shafeey M, Alshehri AG, Alamri S, Shati A, Alrumman S, Alghamdii H (2021) Chitosan functionalized AgNPs for efficient removal of Imidacloprid pesticide through a pressure-free design. International Journal of Biological Macromolecules 168:116-123

Moustafa M, Taha T, Mansy A, Al-Emam A, Alamri S, Alghamdii H, Shati A, Alrumman S, Temerk H, Maghraby T (2020) Dual exploitation of clove powder for bioremediation of heavy metals and decontaminating microbes from wastewater. Desalination and Water Treatment 207:309-320

Mudila H, Prasher P, Kumar A, Sharma M, Verma A, Verma S, Khati B (2021) E-waste and its hazard management by specific microbial bioremediation processes. Microbial Rejuvenation of Polluted Environment: Volume 2:139-166

Munir M, Mardon I, Al-Zuhair S, Shawabkeh A, Saqib N (2019) Plasma gasification of municipal solid waste for waste-to-value processing. Renewable and Sustainable Energy Reviews 116:109461

Mustafa K, Maryam D (2023) Advances in Bioremediation: A Prospective Approach for Green Food Waste Management. FRONTIERS IN CHEMICAL SCIENCES 4 (1):48-67

MWI, 2022. Jordan Water Sector Facts and Figures, Ministry of Water and Irrigation, Amman, Jordan

Naik S, Jujjavarappu SE (2020) Simultaneous bioelectricity generation from cost-effective MFC and water treatment using various wastewater samples. Environmental Science and Pollution Research 27:27383-27393

Narayanan M, Ali SS, El-Sheekh M (2023) A comprehensive review on the potential of microbial enzymes in multipollutant bioremediation: Mechanisms, challenges, and future prospects. Journal of Environmental Management 334:117532

Nivetha N, Srivarshine B, Sowmya B, Rajendiran M, Saravanan P, Rajeshkannan R, Rajasimman M, Pham THT, Shanmugam V, Dragoi E-N (2023) A comprehensive review on bio-stimulation and bio-enhancement towards remediation of heavy metals degeneration. Chemosphere 312:137099

Noman M, Shahid M, Ahmed T, Niazi MBK, Hussain S, Song F, Manzoor I (2020) Use of biogenic copper nanoparticles synthesized from a native Escherichia sp. as photocatalysts for azo dye degradation and treatment of textile effluents. Environmental Pollution 257:113514

Nriagu JO (2019) Encyclopedia of environmental health. Elsevier,

Ojha N, Karn R, Abbas S, Bhugra S Bioremediation of industrial wastewater: A review. In: IOP Conference Series: Earth and Environmental Science, 2021. vol 1. IOP Publishing, p 012012

Ojuederie OB, Chukwuneme CF, Samuel O, Olanrewaju MA, Adegboyega TT, Babalola OO (2021) Contribution of microbial inoculants in sustainable maintenance of human health, including test methods and evaluation of safety of microbial pesticide microorganisms. Biopesticides: Botanicals and Microorganisms for Improving Agriculture and Human Health:207-240

Othman AR, Hasan HA, Muhamad MH, Ismail NI, Abdullah SRS (2021) Microbial degradation of microplastics by enzymatic processes: a review. Environmental Chemistry Letters 19:3057-3073

Palit S, Hussain CM (2020) Functionalization of nanomaterials for industrial applications: recent and future perspectives. In: Handbook of functionalized nanomaterials for industrial applications. Elsevier, pp 3-14

Patel AK, Singhania RR, Albarico FPJB, Pandey A, Chen C-W, Dong C-D (2022) Organic wastes bioremediation and its changing prospects. Science of the Total Environment 824:153889

Patel N, Rai D, Shahane S, Mishra U (2019) Lipases: sources, production, purification, and applications. Recent patents on biotechnology 13 (1):45-56

Principato L, Mattia G, Di Leo A, Pratesi CA (2021) The household wasteful behaviour framework: A systematic review of consumer food waste. Industrial Marketing Management 93:641-649

Qureshi, A.S., 2020. Challenges and prospects of using treated wastewater to manage water scarcity crises in the Gulf Cooperation Council (GCC) countries. Water, 12(7), p.1971.

Raddadi N, Fava F (2019) Biodegradation of oil-based plastics in the environment: Existing knowledge and needs of research and innovation. Science of the Total Environment 679:148-158

Radwan EK, Abdel-Aty AM, El-Wakeel ST, Abdel Ghafar HH (2020) Bioremediation of potentially toxic metal and reactive dye-contaminated water by pristine and modified Chlorella vulgaris. Environmental Science and Pollution Research 27:21777-21789

Rajak R, Mahto RK, Prasad J, Chattopadhyay A (2022) Assessment of bio-medical waste before and during the emergency of novel Coronavirus disease pandemic in

India: A gap analysis. Waste Management & Research 40 (4):470-481

Rastogi M, Nandal M, Khosla B (2020) Microbes as vital additives for solid waste composting. Heliyon 6 (2)

Rautela R, Arya S, Vishwakarma S, Lee J, Kim K-H, Kumar S (2021) E-waste management and its effects on the environment and human health. Science of the Total Environment 773:145623

Rawat S, Verma L, Singh J (2020) Environmental hazards and management of E-waste. Environmental Concerns and Sustainable Development: Volume 2: Biodiversity, Soil and Waste Management:381-398

Ren B, Wang T, Zhao Y (2021) Two-stage hybrid constructed wetland-microbial fuel cells for swine wastewater treatment and bioenergy generation. Chemosphere 268:128803

Ru J, Huo Y, Yang Y (2020) Microbial degradation and valorization of plastic wastes. Frontiers in Microbiology 11:442

Sadeghabad MS, Bahaloo-Horeh N, Mousavi SM (2019) Using bacterial culture supernatant for extraction of manganese and zinc from waste alkaline button-cell batteries. Hydrometallurgy 188:81-91

Saeed MU, Hussain N, Sumrin A, Shahbaz A, Noor S, Bilal M, Aleya L, Iqbal HM (2022) Microbial bioremediation strategies with wastewater treatment potentialities–A review. Science of the total environment 818:151754

Safwat SM, Matta ME (2021) Environmental applications of Effective Microorganisms: a review of current knowledge and recommendations for future directions. Journal of Engineering and Applied Science 68 (1):1-12

Saha L, Tiwari J, Bauddh K, Ma Y (2021) Recent developments in microbe–plant-based bioremediation for tackling heavy metal-polluted soils. Frontiers in Microbiology 12:731723

Sakshi K, Bharadvaja N (2023) Nanotechnology-Based Solutions for Wastewater Treatment. In: Biorefinery for Water and Wastewater Treatment. Springer, pp 71-88

Saleh AK, El-Gendi H, Ray JB, Taha TH (2021) A low-cost effective media from starch kitchen waste for bacterial cellulose production and its application as simultaneous absorbance for methylene blue dye removal. Biomass Conversion and Biorefinery:1-13

Saranya P, Selvi P, Sekaran G (2019) Integrated thermophilic enzyme-immobilized reactor and high-rate biological reactors for treatment of palm oil-containing wastewater without sludge production. Bioprocess and biosystems engineering 42:1053-1064

Saravanan A, Kumar PS, Jeevanantham S, Anubha M, Jayashree S (2022) Degradation of toxic agrochemicals and pharmaceutical pollutants: Effective and alternative approaches toward photocatalysis. Environmental Pollution 298:118844

Saravanan A, Kumar PS, Vo D-VN, Jeevanantham S, Karishma S, Yaashikaa P (2021) A review on catalytic-enzyme degradation of toxic environmental pollutants: Microbial enzymes. Journal of Hazardous Materials 419:126451

Saxena G, Kishor R, Bharagava RN (2020) Application of microbial enzymes in degradation and detoxification of organic and inorganic pollutants. Bioremediation of Industrial Waste for Environmental Safety: Volume I: Industrial Waste and Its Management:41-51

Sayara T, Basheer-Salimia R, Hawamde F, Sánchez A (2020) Recycling of organic wastes through composting: Process performance and compost application in agriculture. Agronomy 10 (11):1838

Sayed K, Baloo L, Sharma NK (2021) Bioremediation of total petroleum hydrocarbons (TPH) by bioaugmentation and biostimulation in water with floating oil spill containment booms as bioreactor basin. International Journal of Environmental Research and Public Health 18 (5):2226

Sedlakova-Kadukova J, Kopcakova A, Gresakova L, Godany A, Pristas P (2019) Bioaccumulation and biosorption of zinc by a novel Streptomyces K11 strain isolated from highly alkaline aluminium brown mud disposal site. Ecotoxicology and Environmental Safety 167:204-211

SERIES, L. P. 2015. *Riyadh Bioremediation Facility* [Online]. Available: https://www.landscapeperformance.org/case-study-briefs/riyadh-bioremediation-facility#project-team [Accessed 24-11-2024].

Shanmuganathan R, Karuppusamy I, Saravanan M, Muthukumar H, Ponnuchamy K, Ramkumar VS, Pugazhendhi A (2019) Synthesis of silver nanoparticles and their biomedical applications-a comprehensive review. Current pharmaceutical design 25 (24):2650-2660

Sharada HM, Abdel-Halim SA, Hafez MA, Elbarbary TA, Abdel-Fatah Y, Ibrahim IA (2021) Bioleaching of Copper from Electronic Waste Using Aspergillus niger.

Sharma M, Nandy A, Taylor N, Venkatesan SV, Kollath VO, Karan K, Thangadurai V, Tsesmetzis N, Gieg LM (2020) Bioelectrochemical remediation of phenanthrene in a microbial fuel cell using an anaerobic consortium enriched from a hydrocarbon-contaminated site. Journal of hazardous materials 389:121845

Sharma P, Dutta D, Udayan A, Nadda AK, Lam SS, Kumar S (2022) Role of microbes in bioaccumulation of heavy metals in municipal solid waste: Impacts on plant and human being. Environmental Pollution 305:119248

Singh D, Gupta N (2020) Microbial laccase: a robust enzyme and its industrial applications. Biologia 75:1183-1193

Singh N, Duan H, Tang Y (2020a) Toxicity evaluation of E-waste plastics and potential repercussions for human health. Environment international 137:105559

Singh P, Singh VK, Singh R, Borthakur A, Madhav S, Ahamad A, Kumar A, Pal DB, Tiwary D, Mishra P (2020b) Bioremediation: a sustainable approach for management of environmental contaminants. In: Abatement of environmental pollutants. Elsevier, pp 1-23

Skariyachan S, Taskeen N, Kishore AP, Krishna BV (2022) Recent advances in plastic degradation–From microbial consortia-based methods to data sciences and computational biology driven approaches. Journal of Hazardous Materials 426:128086

Soliman N, Moustafa A (2020) Industrial solid waste for heavy metals adsorption features and challenges; a review. Journal of Materials Research and Technology 9 (5):10235-10253

Sonawane JM, Rai AK, Sharma M, Tripathi M, Prasad R (2022) Microbial biofilms: Recent advances and progress in environmental bioremediation. Science of The Total Environment 824:153843

Srichandan H, Mohapatra RK, Parhi PK, Mishra S (2019) Bioleaching approach for extraction of metal values from secondary solid wastes: a critical review. Hydrometallurgy 189:105122

Srivastav AL, Ranjan M (2020) Inorganic water pollutants. In: Inorganic Pollutants in Water. Elsevier, pp 1-15

Srivastava S, Kumar M (2019) Biodegradation of polycyclic aromatic hydrocarbons (PAHs): a sustainable approach. Sustainable green technologies for environmental management:111-139

Strokal M, Spanier JE, Kroeze C, Koelmans AA, Flörke M, Franssen W, Hofstra N, Langan S, Tang T, van Vliet MT (2019) Global multi-pollutant modelling of water quality: scientific challenges and future directions. Current opinion in environmental sustainability 36:116-125

Sufficiency E, Qamar SA, Ferreira LFR, Franco M, Iqbal HM, Bilal M (2022) Emerging biotechnological strategies for food waste management: A green leap towards achieving high-value products and environmental abatement. Energy Nexus 6:100077

Sun W, Cheng K, Sun KY, Ma X (2021) Microbially mediated remediation of contaminated sediments by heavy metals: A critical review. Current Pollution Reports 7:201-212

Syed Z, Sogani M, Dongre A, Kumar A, Sonu K, Sharma G, Gupta AB (2021) Bioelectrochemical systems for environmental remediation of estrogens: A review and way forward. Science of the Total Environment 780:146544

Syed Z, Sonu K, Sogani M (2022) Cattle manure management using microbial fuel cells for green energy generation. Biofuels, Bioproducts and Biorefining 16 (2):460-470

Taha RH, Taha TH, Abu-Saied M, Mansy A, Elsherif MA (2022) Maximization of the bioethanol concentration produced through the cardboard waste fermentation by using ethylenediamine-modifying poly (acrylonitrile-co-methyl acrylate) membrane. Biomass Conversion and Biorefinery:1-19

Taha RH, Taha TH, Elsherif MA, Mansy A (2021) Successive Application of Physicochemical and Enzymatic Treatments of Office Paper Waste for the Production of Bioethanol with Possible Using of Carbon Dioxide as an Indicator for the Determination of the Bioethanol Concentration. Journal of Biobased Materials and Bioenergy 15 (6):790-798

Taha TH, Abu-Saied M, Elnouby MS, Hashem M, Alamri S, Mostafa Y (2019) Designing of pressure-free filtration system integrating polyvinyl alcohol/chitosan-silver nanoparticle membrane for purification of microbe-containing water. Water Supply 19 (8):2443-2452

Tahaa TH, Mansyb A, Youssifc AM, Alamrid S, Moustafad M (2021) Speeding up the successive clarification and bioremediation processes of anthracene-containing water using graphite/bacteria integrated columns. Desalination and Water Treatment 223:154-166

Taherzadeh M, Bolton K, Wong J, Pandey A (2019) Sustainable resource recovery and zero waste approaches. Elsevier,

Thakur M, Medintz IL, Walper SA (2019) Enzymatic bioremediation of organophosphate compounds—progress and remaining challenges. Frontiers in bioengineering and biotechnology 7:289

Thapa S, Li H, OHair J, Bhatti S, Chen F-C, Nasr KA, Johnson T, Zhou S (2019) Biochemical characteristics of microbial enzymes and their significance from industrial perspectives. Molecular biotechnology 61:579-601

Tian Q, Dou X, Huang L, Wang L, Meng D, Zhai L, Shen Y, You C, Guan Z, Liao X (2020) Characterization of a robust cold-adapted and thermostable laccase from Pycnoporus sp. SYBC-L10 with a strong ability for the degradation of tetracycline and oxytetracycline by laccase-mediated oxidation. Journal of hazardous materials 382:121084

Torres Munguía JA, Badarau FC, Díaz Pavez LR, Martínez-Zarzoso I, Wacker KM (2022) A global dataset of pandemic-and epidemic-prone disease outbreaks. Scientific data 9 (1):683

Tournier V, Topham C, Gilles A, David B, Folgoas C, Moya-Leclair E, Kamionka E, Desrousseaux M-L, Texier H, Gavalda S (2020) An engineered PET depolymerase to break down and recycle plastic bottles. Nature 580 (7802):216-219

Tran TTT, Pham HK, Nguyen HM (2020) Assessing the current status of rural domestic solid waste management in Nam Dinh province. J Min Earth Sci 61 (6):82-89

Tripathi S, Sharma P, Chandra R (2021) Degradation of organometallic pollutants of distillery wastewater by autochthonous bacterial community in biostimulation and bioaugmentation process. Bioresource Technology 338:125518

Ukaogo PO, Ewuzie U, Onwuka CV (2020) Environmental pollution: causes, effects, and the remedies. In: Microorganisms for sustainable environment and health. Elsevier, pp 419-429

Ul-Abdin Z, Anwar W, Khitab A (2022) Microbiologically induced deterioration of concrete. In: Biodegradation and Biodeterioration at the Nanoscale. Elsevier, pp 389-403

Unuofin JO, Okoh AI, Nwodo UU (2019) Aptitude of oxidative enzymes for treatment of wastewater pollutants: a laccase perspective. Molecules 24 (11):2064

Valles M, Kamaruddin AF, Wong LS, Blanford CF (2020) Inhibition in multicopper oxidases: a critical review. Catalysis Science & Technology 10 (16):5386-5410

Van Geffen L, van Herpen E, van Trijp H (2020) Household Food waste—How to avoid it? An integrative review. Food waste management: Solving the wicked problem:27-55

Verma S, Bhatt P, Verma A, Mudila H, Prasher P, Rene ER (2021) Microbial technologies for heavy metal remediation: effect of process conditions and current practices. Clean Technologies and Environmental Policy:1-23

Verma S, Kuila A (2019) Bioremediation of heavy metals by microbial process. Environmental Technology & Innovation 14:100369

Wang B, Ma J, Zhang L, Su Y, Xie Y, Ahmad Z, Xie B (2021) The synergistic strategy and microbial ecology of the anaerobic co-digestion of food waste under the regulation of domestic garbage classification in China. Science of The Total Environment 765:144632

Wang L, Liu Y, Shu X, Lu S, Xie X, Shi Q (2019) Complexation and conformation of lead ion with poly-γ-glutamic acid in soluble state. Plos one 14 (9):e0218742

Wang Y, Tam NF (2019) Microbial Remediation of organic pollutants. In: World Seas: An Environmental Evaluation. Elsevier, pp 283-303

Wasewar KL, Singh S, Kansal SK (2020) Process intensification of treatment of inorganic water pollutants. In: Inorganic pollutants in water. Elsevier, pp 245-271

Wei Y, Cui M, Ye Z, Guo Q (2021) Environmental challenges from the increasing medical waste since SARS outbreak. Journal of cleaner production 291:125246

Xie Q, Liu N, Lin D, Qu R, Zhou Q, Ge F (2020) The complexation with proteins in extracellular polymeric substances alleviates the toxicity of Cd (II) to Chlorella vulgaris. Environmental Pollution 263:114102

Xue S-W, Tian Y-X, Pan J-C, Liu Y-N, Ma Y-L (2021) Binding interaction of a ring-hydroxylating dioxygenase with fluoranthene in Pseudomonas aeruginosa DN1. Scientific Reports 11 (1):21317

Yadav AN, Suyal DC, Kour D, Rajput VD, Rastegari AA, Singh J (2022) Bioremediation and waste management for environmental sustainability. Journal of Applied Biology and Biotechnology 10 (2):1-5

Yang Y, Bao W, Xie GH (2019) Estimate of restaurant food waste and its biogas production potential in China. Journal of cleaner production 211:309-320

Yap PL, Nine MJ, Hassan K, Tung TT, Tran DN, Losic D (2021) Graphene-based sorbents for multipollutants removal in water: a review of recent progress. Advanced Functional Materials 31 (9):2007356

Ye J, Song Y, Liu Y, Zhong Y (2022) Assessment of medical waste generation, associated environmental impact, and management issues after the outbreak of COVID-19: A case study of the Hubei Province in China. PloS one 17 (1):e0259207

Ye L, Zhong W, Zhang M, Jing C (2021) New mobilization pathway of antimonite: Thiolation and oxidation by dissimilatory metal-reducing bacteria via elemental sulfur respiration. Environmental Science & Technology 56 (1):652-659

Ye X, Peng T, Feng J, Yang Q, Pratush A, Xiong G, Huang T, Hu Z (2019) A novel dehydrogenase 17β-HSDx from Rhodococcus sp. P14 with potential application in bioremediation of steroids contaminated environment. Journal of Hazardous Materials 362:170-177

Yusoff DF, Raja Abd Rahman RNZ, Masomian M, Ali MSM, Leow TC (2020) Newly isolated alkane hydroxylase and lipase producing Geobacillus and Anoxybacillus species involved in crude oil degradation. Catalysts 10 (8):851

Zabermawi NM, Alsulaimany FA, El-Saadony MT, El-Tarabily KA (2022a) New eco-friendly trends to produce biofuel and bioenergy from microorganisms: An updated review. Saudi Journal of Biological Sciences

Zabermawi NMO, Alyhaiby AH, El-Bestawy EA (2022b) Microbiological analysis and bioremediation bioassay for characterization of industrial effluent. Scientific Reports 12 (1):18889

Zamri MLA, Makhtar SMZ, Sobri MFM, Makhtar MMZ Microbial Fuel Cell as New Renewable Energy for Simultaneous Waste Bioremediation and Energy Recovery. In: IOP Conference Series: Earth and Environmental Science, 2023. vol 1. IOP Publishing, p 012035

Zhang H, Yuan X, Xiong T, Wang H, Jiang L (2020a) Bioremediation of co-contaminated soil with heavy metals and pesticides: Influence factors, mechanisms and evaluation methods. Chemical Engineering Journal 398:125657

Zhang K, Yang W, Liu Y, Zhang K, Chen Y, Yin X (2020b) Laccase immobilized on chitosan-coated Fe_3O_4 nanoparticles as reusable biocatalyst for degradation of chlorophenol. Journal of Molecular Structure 1220:128769

Zhang Y, Cui Y, Chen P, Liu S, Zhou N, Ding K, Fan L, Peng P, Min M, Cheng Y (2019a) Gasification technologies and their energy potentials. In: Sustainable resource recovery and zero waste approaches. Elsevier, pp 193-206

Zhang Y, Liu M, Zhou M, Yang H, Liang L, Gu T (2019b) Microbial fuel cell hybrid systems for wastewater treatment and bioenergy production: synergistic effects, mechanisms and challenges. Renewable and Sustainable Energy Reviews 103:13-29

Zhao F, Wang S (2019) Bioleaching of electronic waste using extreme acidophiles. In: Electronic Waste Management and Treatment Technology. Elsevier, pp 153-174

Zheng Y-M, Xi B-D, Shan G-C, Yu M-D, Cui J, Wei K-H, Liu H-B, He X-S (2021) High proportions of petroleum loss ascribed to volatilization rather than to microbial degradation in greenhouse-enhanced biopile. Journal of Cleaner Production 303:127084

Zhou M-H, Shen S-L, Xu Y-S, Zhou A-N (2019) New policy and implementation of municipal solid waste classification in Shanghai, China. International journal of environmental research and public health 16 (17):3099

Zhou Y, Sun F, Wu X, Cao S, Guo X, Wang Q, Wang Y, Ji R (2022) Formation and nature of non-extractable residues of emerging organic contaminants in humic acids catalyzed by laccase. Science of The Total Environment 829:154300

Zouboulis AI, Moussas PA, Psaltou SG (2019) Groundwater and soil pollution: bioremediation.

ABDULMOHSEN, K. A., ALIMI, F. R., MECHI, L., KAA, A. A., MOHAMMAD, A. E. & KHAN, M. W. A. 2023. Optimization of Pseudomonas aeruginosa isolated for bioremediation from Ha'il region of Saudi Arabia. *Bioinformation,* 19**,** 893.

AL-DHABI, N. A., ESMAIL, G. A. & VALAN ARASU, M. 2020. Enhanced production of biosurfactant from Bacillus subtilis strain Al-Dhabi-130 under solid-state fermentation using date molasses from Saudi Arabia for bioremediation of crude-oil-contaminated soils.

International Journal of Environmental Research and Public Health, 17, 8446.

HASHEM, A. 2007. Bioremediation of petroleum contaminated soils in the Arabian Gulf region: a review. *Science,* 19.

JONES, D. A., HAYES, M., KRUPP, F., SABATINI, G., WATT, I. & WEISHAR, L. 2008. The impact of the Gulf War (1990–91) oil release upon the intertidal Gulf coast line of Saudi Arabia and subsequent recovery. *Protecting the Gulf's marine ecosystems from pollution.* Springer.

MOUSTAFA, A. 2016. Bioremediation of oil spill in Kingdom of Saudi Arabia by using fungi isolated from polluted soils. *International Journal of Current Microbiology and Applied Sciences,* 5, 680-91.

SERIES, L. P. 2015. *Riyadh Bioremediation Facility* [Online]. Available: https://www.landscapeperformance.org/case-study-briefs/riyadh-bioremediation-facility#project-team [Accessed 24-11-2024].

MWI, 2022. Jordan Water Sector Facts and Figures, Ministry of Water and Irrigation, Amman, Jordan

Qureshi, A.S., 2020. Challenges and prospects of using treated wastewater to manage water scarcity crises in the Gulf Cooperation Council (GCC) countries. Water, 12(7), p.1971.

Bahadir, M., Aydin, M.E., Aydin, S., Beduk, F. and Batarseh, M., 2016. Wastewater reuse in Middle East Countries—A review of prospects and challenges. Fresenius Environmental Bulletin, 25(5), pp.1284-1304.

Belhassan, K., 2022. Managing drought and water stress in Northern Africa. In Arid Environment-Perspectives, Challenges and Management. IntechOpen.

Gado, T.A. and El-Agha, D.E., 2020. Feasibility of rainwater harvesting for sustainable water management in urban areas of Egypt. Environmental Science and Pollution Research, 27(26), pp.32304-32317.

Alsadey, S. and Mansour, O., 2020. Wastewater treatment plants in Libya: Challenges and future prospects. International Journal of Environmental Planning and Management, 6(3), pp.76-80.

Chaouali, S., Dos Muchangos, L.S., Ito, L. and Tokai, A., 2024. Assessment of the Environmental Impacts of Wastewater Treatment in Tunisia. Journal of Water and Environment Technology, 22(2), pp.61-74.

www.ingramcontent.com/pod-product-compliance
Lightning Source LLC
Chambersburg PA
CBHW071459220526
45472CB00003B/853